LANDFILL CLOSURES

...ENVIRONMENTAL PROTECTION and LAND RECOVERY

Proceedings of sessions sponsored by the
Environmental Geotechnics Committee of the
Geotechnical Engineering Division and the
Solid Waste Engineering Committee of the
Environmental Engineering Division of the
American Society of Civil Engineers
in conjunction with the ASCE Convention in
San Diego, California,

Edited by R. Jeffrey D

Geotechnical Special Publication No. 53

Published by the
American Society of Civil Engineers
345 East 47th Street
New York, New York 10017-2398

ABSTRACT

This proceedings, Landfill Closures - Environmental Protection and Land Recovery, consists of papers presented at the ASCE Annual Convention and Exposition held in San Diego, California, October 23-27, 1995. With increased closure and post-closure development of municipal solid waste landfills, there has been much study and design related to topics such as: 1.) Suitable landfill cover design, 2.) dry landfills versus wet ones, 3.) leachates, 4.) landfill gas, and 5.) condensate. These and other subjects are discussed in this proceedings.

Library of Congress Cataloging-in-Publication Data

Landfill closures— environmental protection and land recovery : pro
    ceedings of sessions sponsored by the Environmental
    Geotechnics Committee of the Geotechnical Engineering Division
    and the Solid Waste Engineering Committee of the Environmental
    Engineering Division of the American Society of Civil Engineers in
    conjunction with the ASCE convention in San Diego, California,
    October 23-27, 1995 / edited by R. Jeffrey Dunn and Udai P.
    Singh
        p.   cm. — (Geotechnical special publication ; no. 53)
        ISBN 0-7844-0119-5
        1. Sanitary landfill closures—Congresses.
    I. Dunn, R. Jeffrey. II. Singh, Udai O. III. American Society
of Civil Engineers. Geotechnical Engineering Division.
Environmental Geotechnics Committee. IV. American Society of
Civil Engineers. Environmental Engineering Division. Solid
Waste Engineering Committee.V. Series.
TD795.7.L342 1995   95-40547
363.72'8—dc20   CIP

Cover photos and graphics: Babylon Landfill Closure,
    Babylon, New York
Photo by Kenneth Cargill, Graphics by Ann Taylor, GeoSyntec
    Consultants, Atlanta, Georgia

GEOTECHNICAL SPECIAL PUBLICATIONS

1) TERZAGHI LECTURES
2) GEOTECHNICAL ASPECTS OF STIFF AND HARD CLAYS
3) LANDSLIDE DAMS: PROCESSES RISK, AND MITIGATION
4) TIEBACKS FOR BULKHEADS
5) SETTLEMENT OF SHALLOW FOUNDATION ON COHESIONLESS
   SOILS: DESIGN AND PERFORMANCE
6) USE OF IN SITU TESTS IN GEOTECHNICAL ENGINEERING
7) TIMBER BULKHEADS
8) FOUNDATIONS FOR TRANSMISSION LINE TOWERS
9) FOUNDATIONS AND EXCAVATIONS IN DECOMPOSED ROCK OF
   THE PIEDMONT PROVINCE
10) ENGINEERING ASPECTS OF SOIL EROSION DISPERSIVE CLAYS AND LOESS
11) DYNAMIC RESPONSE OF PILE FOUNDATIONS— EXPERIMENT,
    ANALYSIS AND OBSERVATION
12) SOIL IMPROVEMENT - A TEN YEAR UPDATE
13) GEOTECHNICAL PRACTICE FOR SOLID WASTE DISPOSAL '87
14) GEOTECHNICAL ASPECTS OF KARST TERRIANS
15) MEASURED PERFORMANCE SHALLOW FOUNDATIONS
16) SPECIAL TOPICS IN FOUNDATIONS
17) SOIL PROPERTIES EVALUATION FROM CENTRIFUGAL MODELS
18) GEOSYNTHETICS FOR SOIL IMPROVEMENT
19) MINE INDUCED SUBSIDENCE: EFFECTS ON ENGINEERED STRUCTURES
20) EARTHQUAKE ENGINEERING & SOIL DYNAMICS (II)
21) HYDRAULIC FILL STRUCTURES
22) FOUNDATION ENGINEERING
23) PREDICTED AND OBSERVED AXIAL BEHAVIOR OF PILES
24) RESILIENT MODULI OF SOILS: LABORATORY CONDITIONS
25) DESIGN AND PERFORMANCE OF EARTH RETAINING STRUCTURES
26) WASTE CONTAINMENT SYSTEMS; CONSTRUCTION, REGULATION, AND
    PERFORMANCE
27) GEOTECHNICAL ENGINEERING CONGRESS
28) DETECTION OF AND CONSTRUCTION AT THE SOIL/ROCK INTERFACE
29) RECENT ADVANCES IN INSTRUMENTATION, DATA ACQUISITION AND
    TESTING IN SOIL DYNAMICS
30) GROUTING, SOIL IMPROVEMENT AND GEOSYNTHETICS
31) STABILITY AND PERFORMANCE OF SLOPES AND EMBANKMENTS II
    (A 25 YEAR PERSPECTIVE)
32) EMBANKMENT DAMS-JAMES L. SHERARD CONTRIBUTIONS
33) EXCAVATION AND SUPPORT FOR THE URBAN INFRASTRUCTURE
34) PILES UNDER DYNAMIC LOADS
35) GEOTECHNICAL PRACTICE IN DAM REHABILITATION
36) FLY ASH FOR SOIL IMPROVEMENT
37) ADVANCES IN SITE CHARACTERIZATION: DATA ACQUISITION,
    DATA MANAGEMENT AND DATA INTERPRETATION
38) DESIGN AND PERFORMANCE OF DEEP FOUNDATIONS: PILES AND
    PIERS IN SOIL AND SOFT ROCK
39) UNSATURATED SOILS
40) VERTICAL AND HORIZONTAL DEFORMATIONS OF FOUNDATIONS AND
    EMBANKMENTS
41) PREDICTED AND MEASURED BEHAVIOR OF FIVE SPREAD FOOTINGS ON SAND
42) SERVICEABILITY OF EARTH RETAINING STRUCTURES
43) FRACTURE MECHANICS APPLIED TO GEOTECHNICAL ENGINEERING
44) GROUND FAILURES UNDER SEISMIC CONDITIONS
45) IN-SITU DEEP SOIL IMPROVEMENT
46) GEOENVIRONMENT 2000
47) GEO-ENVIRONMENTAL ISSUES FACING THE AMERICAS
48) SOIL SUCTION APPLICATIONS IN GEOTECHNICAL ENGINEERING
49) SOIL IMPROVEMENT FOR EARTHQUAKE HAZARD MITIGATION
50) FOUNDATION UPGRADING AND REPAIR FOR INFRASTRUCTURE IMPROVEMENT
51) PERFORMANCE OF DEEP FOUNDATIONS UNDER SEISMIC LOADING
52) LANDSLIDES UNDER STATIC AND DYNAMIC CONDITIONS - ANALYSIS,
    MONITORING, AND MITIGATION
53) LANDFILL CLOSURES - ENVIRONMENTAL PROTECTION AND LAND RECOVERY
54) EARTHQUAKE DESIGN AND PERFORMANCE OF SOLID WASTE LANDFILLS
55) EARTHQUAKE - INDUCED MOVEMENTS AND SEISMIC REMEDIATION OF
    EXISTING FOUNDATIONS AND ABUTMENTS
56) STATIC AND DYNAMIC PROPERTIES OF GRAVELLY SOILS

PREFACE

Municipal solid waste (MSW) landfills are closing at a rapid rate as they reach capacity or in response to costs of operations. These increased operations costs are largely due to compliance with ever more complex and rigorous regulations governing MSW landfills at the local, state, and federal level. The days of the "mom and pop" landfill run by a small business, or of the small rural landfill are drawing to a close. These facilities are now being replaced by transfer stations, where wastes are collected and transported to medium to large sized regional facilities operated by public agencies and private businesses. At the same time, closed landfills, mainly in urban areas, are being developed into a variety of new uses from parks to commercial facilities as land becomes scarce and expensive. Increased closure and post-closure development of MSW landfills has led to much engineering study and design related to the topics of suitable landfill cover design, the merits of dry landfills versus wet ones where waste decomposition is promoted, as well as the need for improved practices for handling and treatment of leachate, landfill gas, and condensate. Observation of this engineering activity prompted the Environmental Geotechnics Committee of the Geotechnical Engineering Division and the Solid Waste Engineering Committee of the Environmental Engineering Division to jointly sponsor three specialty technical sessions at the October 1995 ASCE National Convention in San Diego, California, on the topic of landfill closures. Speakers for the sessions were selected and invited by the sponsoring committees.

In addition to the three specialty sessions, the committees decided to prepare a proceedings volume, for publication by ASCE, of papers prepared by the invited speakers and additional papers selected by the proceedings editors from authors who submitted abstracts in response to a call for papers published in ASCE News. All published papers were peer-reviewed for technical quality and content before being accepted for publication. The standards of review were essentially the same as those of the ASCE *Journal of Geotechnical Engineering* and *Journal of Environmental Engineering*. Each paper received two positive reviews before being accepted and was revised to conform to any mandatory revisions of the reviewers. It should also be noted that all papers in this volume are eligible for discussion in the *Journal of Geotechnical Engineering* or *Journal of Environmental Engineering* and for ASCE awards.

The editors would like to extend their thanks to Hari Sharma and Russ Scharlin, members of the sessions planning committee, who greatly assisted with the planning of the specialty sessions, and to the members of the Environmental Geotechnics Committee and Solid Waste Engineering Committee who provided valuable input during the planning of the sessions. They would also like to express their sincere gratitude to the many reviewers from these two committees and the engineering community who completed the peer reviews, in many cases in a very limited timeframe.

Finally, the editors extend their thanks to Shiela Menaker and the staff at ASCE who managed the assembly and printing of this proceeding volume and provided guidance during its preparation.

R. Jeffrey Dunn
Principal
GeoSyntec Consultants
Walnut Creek, California

Udai P. Singh
Senior Environmental Engineer
CH2M HILL
Oakland, California

CONTENTS

ADDITIONAL PROCEEDINGS PAPERS

SESSION CV98
POST CLOSURE ISSUES AND LAND DEVELOPMENT
Moderator: Hari D. Sharma

INVITED SPEAKER PAPERS

ADDITIONAL PROCEEDINGS PAPERS

Soil Barrier Layers Versus Geosynthetic Barriers in Landfill Cover Systems

David E. Daniel[1]

Abstract

The materials that have traditionally been used for the barrier layer within final covers are, in decreasing order of usage: (1) low-hydraulic-conductivity, clayey, compacted soil; (2) geomembranes; and (3) geosynthetic clay liners (GCLs). Unless the clayey, compacted soil barrier is buried under a thick layer of protective soil or covered with a geomembrane, the compacted soil is likely to desiccate and crack. Large differential settlement from uneven compression of underlying waste is almost certain to produce cracks within a compacted soil barrier in a final cover system. Geomembranes do not suffer as much from these problems, and geosynthetic clay liners are better able to resist damage from freeze-thaw, desiccation, and differential settlement than compacted, clayey soil. It is recommended geomembranes and GCLs be used more frequently and that compacted soil be used less frequently for final covers in which an extremely low-hydraulic-conductivity barrier is desired. However, caution should be exercised for covers with steep slopes due to potential internal shear failure (compacted soil or GCL) or interface shear failure (geomembrane or GCL).

Introduction

Final cover systems are a critical component in landfills. The final cover system separates buried waste or contaminated material from the surface environment, restricts infiltration of water into the waste, and (in some cases) limits release of gas from the waste. If the objective is prevention of pollution to ground water, then one obvious strategy is to minimize the amount of liquid percolating through the final cover (EPA, 1989).

---

[1] Prof. of Civil Engineering, Univ. of Texas, Austin, TX 78712

1

At the present time, most of the engineered covers that are constructed in the U.S. are for municipal solid waste (MSW) landfills. Final covers are also constructed for hazardous waste landfills, non-hazardous industrial waste landfills, disposal facilities for mining and other special wastes, and closure projects for remediation or site restoration activities.

The design of the final cover is complicated by:

• Temperature extremes, possibly including freeze/thaw to significant depths;
• Cyclic wetting and drying of soils;
• Plant roots, burrowing animals, and insects in soil;
• Differential settlement caused by uneven settlement of the underlying waste or foundation soil;
• Down-slope slippage or creep;
• Vehicular movements on roads traversing the cover;
• Wind or water erosion; and
• Deformations caused by earthquakes.

The design of a cover system can be challenging. It is often more difficult to provide an effective hydraulic barrier layer in a cover system than in a liner system because the cover system is challenged in a larger variety of ways by mostly unknown and unquantifiable stresses that do not act on liner systems buried deep beneath the waste.

The purpose of this paper is to contrast three types of low-hydraulic-conductivity barrier materials that might be used in final cover systems for landfills:

• Low-hydraulic-conductivity compacted soil;
• Geomembranes; and
• Geosynthetic clay liners.

Historically, low-hydraulic-conductivity compacted soil has been the material of choice for the barrier layer in landfill closures. In the past decade, geomembranes have been used to a much greater extent, often in combination with compacted soil. Geosynthetic clay liners (GCLs) have received greatly increased use within the past 5 years. Each material has advantages and disadvantages, which are reviewed, analyzed, and contrasted in this paper.

Requirements for Final Cover Systems

Most engineered cover systems are composed of multiple components. As shown in Fig. 1, the components of a cover system can be grouped into five categories: (1) surface layer; (2) protection layer; (3) drainage layer; (4) barrier layer; and (5) gas collection/foundation layer.

Not all components are needed for all final covers. For example, a drainage layer might not be needed for a landfill located in an arid region. Similarly, a gas collection layer may be required at some landfills but not others. Further, some layers may be combined -- for instance, the surface layer and protection layer may be combined into a single layer of cover soil over the surface of the landfill. The materials used in constructing final covers include natural earth materials, geosynthetics, waste materials, blends of natural materials, or blends of earth and geosynthetic materials.

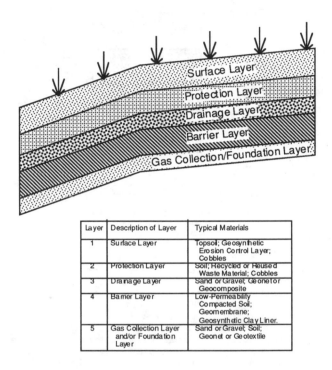

Layer	Description of Layer	Typical Materials
1	Surface Layer	Topsoil; Geosynthetic Erosion Control Layer; Cobbles
2	Protection Layer	Soil; Recycled or Reused Waste Material; Cobbles
3	Drainage Layer	Sand or Gravel; Geonet or Geocomposite
4	Barrier Layer	Low-Permeability Compacted Soil; Geomembrane; Geosynthetic Clay Liner.
5	Gas Collection Layer and/or Foundation Layer	Sand or Gravel; Soil; Geonet or Geotextile

Figure 1.   Five Basic Components of a Final Cover.

In the U.S. requirements for new waste disposal facilities are established by the U.S. Environmental Protection Agency (EPA). Individual states may impose regulations that are more restrictive. The EPA's guidance

on minimum requirements for final cover systems for
hazardous waste landfills are shown in Fig. 2. The EPA
also recommends an alternative final cover, sketched in
Fig. 3, which appears to apply primarily to arid sites.
The EPA's minimum requirements for MSW landfills are
summarized in Fig. 4.

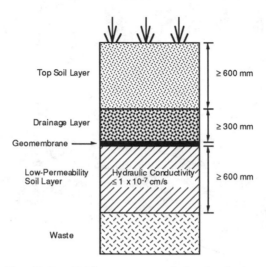

Figure 2. The U.S. EPA's Recommended Final Cover Profile
for Hazardous Waste Landfills (EPA, 1989).

The EPA's requirements for MSW landfill covers are
confusing (Austin, 1992). Regulations state that the
cover must have a hydraulic conductivity that is less than
or equal to the hydraulic conductivity of any bottom liner
system or natural subsoils present, or have a hydraulic
conductivity no greater than 1×10^{-5} cm/s, whichever is
less. This implies that if the bottom liner system
contains a composite geomembrane/clay liner (with the clay
liner having a hydraulic conductivity, k, $\leq 1 \times 10^{-7}$
cm/s), then the final cover should also have a
geomembrane/clay barrier and that the clay barrier should
have k $\leq 1 \times 10^{-7}$ cm/s. However, the EPA's
"clarification" in the June 26, 1992, Federal Register,
notes that if the bottom liner consists of the
prescriptive Subtitle D composite liner with k $\leq 1 \times 10^{-7}$
cm/s, the cover must have a geomembrane overlying 18
inches of 1×10^{-5} cm/s soil. It is not clear why the
hydraulic conductivity can be 1×10^{-5} cm/s, rather than $\leq$
1×10^{-7} cm/s, as would appear to be necessary.

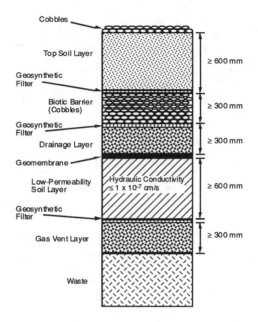

Figure 3. The U.S. EPA's Recommended Alternative Final Cover Profile for Hazardous Waste Landfills (EPA, 1989).

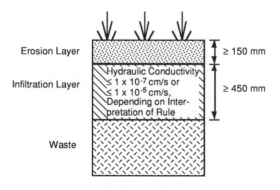

Figure 4. The U.S. EPA's Requirements for Final Cover Profile for Municipal Solid Waste Landfills.

Barrier Layer

     The barrier layer (Fig. 1) is often viewed as the
most critical component of an engineered final cover
system. The barrier layer minimizes percolation of water
through the cover system directly by hydraulic impedance
and indirectly by promoting storage or drainage of water
in the overlying layers, where water is eventually removed
by runoff, evapotranspiration, or internal drainage.

     It is unfortunate that low-hydraulic-conductivity
compacted soil has been so widely and indiscriminately
used in cover systems because there are several problems
which make the long-term performance of a compacted soil
liner questionable. The problems with low-hydraulic-
conductivity compacted soil (clay) liners used in cover
systems include:

 • Clay liners are difficult to compact properly on a
   soft foundation (i.e., waste);
 • Compacted clay will tend to desiccate from above
   and/or below and crack unless protected adequately;
 • Compacted clay is vulnerable to damage from freezing
   and must be protected from freezing by a suitably
   thick layer of cover soil;
 • Differential settlement of underlying compressible
   waste will cause cracking in the compacted clay if
   tensile strains in the clay become excessive; and
 • Compacted clay liners are difficult to repair if they
   are damaged.

These potential problems are explored in further detail in
the subsections that follow.

Field Performance of Barrier Layers

     Several field test plots provide valuable data on the
performance of cover systems under realistic field
conditions. Three cases are reviewed below.

     Omega Hills Landfill. Montgomery and Parsons (1989)
describe field experiments in which three test plots were
constructed and monitored for 3 years at the Omega Hills
Landfill near Milwaukee, Wisconsin. The cross sections
are shown in Fig. 5. In Test Plots 1 and 2, a 1.2-m (4-
ft) thick layer of compacted clay was covered with either
150 mm (6 in) or 450 mm (18 in) of topsoil. The purpose
of Test Plot 3 was to study the use of a capillary break
to prevent desiccation of the underlying 600 mm (2 ft) of
compacted clay. The test plots were installed on the
actual landfill, on a 3(H):1(V) slope. After 3 years,
excavations were made into the cover systems. In all
three plots, the upper 200 to 250 mm of clay was weathered

and blocky. Cracks up to 12 mm wide extended up to 1 m into the clay in Test Plots 1 and 2 (and all the way through the upper clay layer in Test Plot 3). Roots penetrated up to 250 mm into the clay in a continuous mat in all three plots. Some roots extended up to 750 mm into the clay in Test Plots 1 and 2. Neither 150 mm (6 in) nor 450 mm (18 in) of topsoil was enough to protect the underlying clay layer adequately. In Test Plot 3, the lower clay layer was not desiccated.

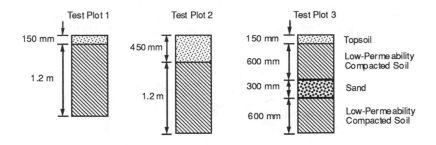

Figure 5. Test Plots at Omega Hills Landfill (Montgomery and Parsons, 1989)

Kettlemen Hills. Corser and Cranston (1991) describe test plots in which a layer of low-hydraulic-conductivity compacted soil (soil-bentonite blend) was covered with either (1) a 600 mm layer of topsoil, (2) an unprotected geomembrane, or (3) a geomembrane overlain by 450 mm of topsoil (Fig. 6). The site was located in a relatively arid part of California at the Kettlemen Hills facility. In less than a year, significant drying and cracking occurred in the plots with soil cover alone and the geomembrane cover alone, but no significant desiccation occurred in the clay covered with both a geomembrane and soil. Only the plot in which the low-hydraulic-conductivity compacted soil liner was covered with a geomembrane and overlying cover soil was the liner protected from desiccation.

Landfill in Hamburg, Germany. Melchior et al. (1994) monitored the field performance of three field plots (each measuring 10 m by 50 m in plan dimension) at a site near Hamburg, Germany, for a 5-year period. The profiles of the plots are shown in Fig. 7. Lysimeters were installed beneath test plots to monitor the rate of water percolation through the test plots. The test plots consisted of 750 mm (30 in) of topsoil overlying a

geotextile filter, which was underlain by a 400-mm (16 in)
thick drainage layer. The barrier layers in the three
test plots were different and consisted of: (1) 600 mm (2
ft) of low-hydraulic-conductivity compacted soil; (2) a
high density polyethylene (HDPE) geomembrane underlain by
600 mm of low-hydraulic-conductivity compacted soil; and
(3) 400 mm (16 in) of low-hydraulic-conductivity compacted
soil. Each test plot was partially located on a 20% slope
and partially located on a 4% slope. The compacted soil
was a glacial till, which was compacted to > 95% of
standard Proctor density on the wet side of optimum. The
mean hydraulic conductivity of the low-hydraulic-
conductivity compacted soil barrier, measured in the
laboratory during construction, was 2.4 x 10^{-8} cm/s.

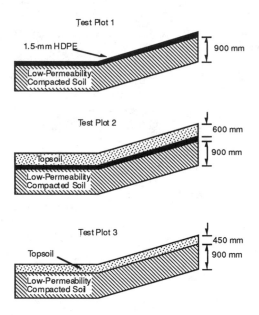

Figure 6. Test Plots at Kettlemen Hills (Corser and
        Cranston, 1991).

        Melchior et al. (1994) monitored the flow rate in
lysimeters for 5 years. Seepage took place through the
barrier layers during most of the year, although in the
summer months flow often stopped, and the gradient

(measured by tensiometers inserted in the soil layers) was upward. The summer of 1992 (the fifth year for the 5-year period described) was an extremely dry summer (actual precipitation not reported).

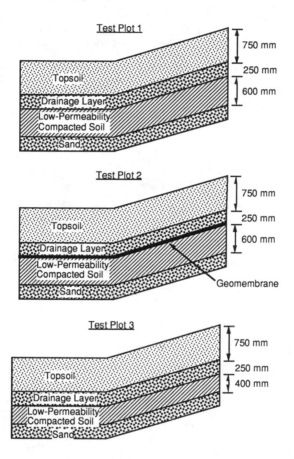

Figure 7. Test Plots at Landfill in Hamburg, Germany (Melchior et al., 1994).

The average annual percolation rates through the barrier layers are summarized in Table 1. (For reference, a flux of 1 x 10^{-7} cm/s corresponds to 32 mm/yr.) The compacted soil liners performed well in the first two years, but then the percolation rate began to increase.

Excavations made into the test plots in 1993 revealed the existence of very small fissures between soil aggregates (about 50 mm in diameter) in the low-hydraulic-conductivity compacted soil barriers that were not protected by a geomembrane. The fissures were visible only because there was shiny free water on the surface of the aggregates (peds) of clay. Melchior et al. report that, "Hypotheses other than shrinkage to explain the formation of the macropores can either be excluded (freezing, penetration of roots, or burrowing by animals), or are not supported by the available data (subsidence)." Plant roots reached the upper part of the barrier layers in 1992. From 1992 on, plant roots contributed directly to the desiccation and formation of soil structure in the sections that lacked a geomembrane.

Table 1.   Annual Percolation of Water through Flat Portions of Barrier Layers for Test Sections in Hamburg, Germany (Melchior et al., 1994).

	Average Annual Percolation (mm/year)		
Year	Plot 1	Plot 2	Plot 3
1988	7.0	3.4	8.4
1989	8.0	0.6	14.2
1990	17.5	0.4	31.0
1991	8.6	0.5	32.3
1992	102.5	0.8	101.2
TOTAL	143.6	5.7	187.1

The seepage through the geomembrane/clay composite liner (Plot 2) was zero in the wet months of the year, when the head on the composite liner was greatest, and was measurable during the dry summer months. The flow collected in the lysimeters was not associated with leaks in the geomembrane but was thought to be related to a slight desiccation of the clay from dry conditions beneath the liner during summer months and may have been thermally driven.

The 5-year totals are shown in Fig. 8. The HDPE/compacted soil composite liner allowed approximately 50 times less percolation of water through the barrier compared to the low-hydraulic-conductivity compacted soil liners.

Summary. The case studies illustrate that the best and perhaps only practical way to protect a low-hydraulic-conductivity compacted soil liner from desiccation from

the surface is to cover the soil liner with both a
geomembrane and a layer of cover soil. To provide less
protection is inappropriate if the designer's intention is
for the low-hydraulic-conductivity compacted soil liner to
remain intact and not crack. A geomembrane may not be
necessary if an extremely thick layer of cover soil is
used, but the layer would have to be so thick that for
most projects soil alone would not be practical.

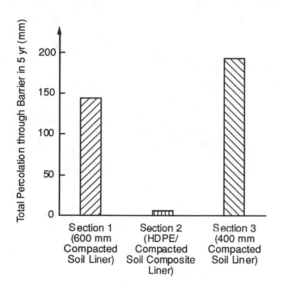

Figure 8. Total Percolation through Barrier Layers at Site
in Hamburg, Germany (Melchior, 1994).

Differential Settlement

LaGatta has summarized available data on the tensile
strains at failure in compacted clay, and these data are
shown in Table 2. The published data indicate that most
compacted clays cannot withstand tensile strains greater
than about 0.1 to 1.0% without cracking. The question,
then, is how likely it is that tensile strains greater
than 0.1 to 1.0% will develop in landfill covers.

It is convenient to define *distortion* as the
differential settlement Δ that occurs over a distance L
(see Fig. 9). Distortion in a cover stretches the barrier

layer and, as a result, produces tensile strains in the
cover components. Murphy and Gilbert (1985) computed the
relationship between distortion and tensile strain for
cover components. The maximum Δ/L associated with a
tensile strain of 0.1% to 1% is 0.05 to 0.1.

Table 2.   Tensile Strain at Failure of Compacted Soils
            (LaGatta, 1992).

Reference	Type or Source of Soil	Water Content (%)	Plasticity Index (%)	Tensile Strain at Failure (%)
Tschebotar-ioff and DePhilippe (1953)	Natural Clayey Soil	19.9	7	0.80
	Bentonite (Montmoril-lonite)	101	487	3.4
	Illite	31.5	34	0.84
	Kaolinite	37.6	38	0.16
Leonards and Narain (1963)	Portland Dam Portland, CO	16.3	8	0.14
	Rector Creek Dam Napa, CA	19.8	16	0.16
	Woodcrest Dam Riverside, CA	10.2	Non-plastic	0.18%
	Shell Oil Dam Ventura, CA	11.2	Non-plastic	0.07%
	Willard Test Dam Embankment, UT	16.4	11	0.20%

Jessberger and Stone (1991) performed centrifuge
tests on 35- to 45-mm-thick clay liners (kaolinite and a
sand-bentonite mixture) that would be used for cover
systems. A "trap door" beneath the liner was deformed to
produce settlement like that which would be expected in a
landfill. Flow rates through the soil were measured as a
function of the distortion in the liner. It was found
that tensile cracking occurred in the liners when
distortion reached $\Delta/L \approx 0.1$ (tensile strain $\approx$ 0.5%). For
kaolinite k increased from the initial value of 1×10^{-7}
cm/s to about 1×10^{-3} cm/s. The experiments reported by
Jessberger and Stone support the findings from Murphy and
Gilbert (1985), i.e., low-hydraulic-conductivity compacted
soil liners in cover systems cannot withstand Δ/L's
greater than approximately 0.05 to 0.1 without cracking.

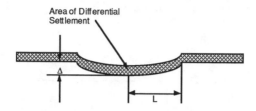

Figure 9. Definition of Distortion, Δ/L.

What does this level of distortion mean in practical terms? Suppose that one observes a circular crater in a landfill cover with a diameter of 5 m (16 ft). What is the maximum settlement at the center of the crater before significant cracking would be expected in the liner? The horizontal distance from the edge to center of the crater, L, is 2.5 m (8 ft). If a maximum Δ/L of 0.05 to 0.1 is multiplied times L, the resulting maximum allowable settlement (Δ) is 0.125 to 0.25 m (5 to 10 in). It is the authors' experience that many, if not all, covers for municipal solid waste landfills have areas with distortions of this magnitude or larger. In such cases, it may be pointless to waste time and money compacting an extremely impermeable layer of compacted clay if the layer is destined to suffer tension cracks and to lose its initially low hydraulic conductivity.

In contrast to soil, geomembranes can withstand large tensile strains, even when stressed three dimensionally. Tensile stains of 20 to 100%, depending on the material, can be taken without rupture in geomembranes that are subjected to 3-dimensional strains, such as would be experienced in landfill covers (Koerner, 1994). The maximum tensile strain that a geomembrane can withstand is at least an order of magnitude greater than for a compacted clay. Indeed, a geomembrane suffers from few of the problems experienced by compacted clay in a cover system; the geomembrane can withstand large differential settlement and is not vulnerable to damage from desiccation or freeze-thaw.

LaGatta (1992) has studied the response of GCLs to differential settlement and has found that, while results vary from one GCL to another, GCLs can typically withstand tensile strains of at least 1% to 10% without loss of low hydraulic conductivity. One example of the relationship between hydraulic conductivity and tensile strain for a

geotextile-encased, needle-punched GCL is shown in Fig.
10.

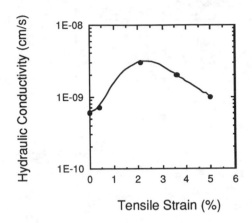

Figure 10. Relationship between Hydraulic Conductivity and
            Tensile Strain in a GCL (LaGatta, 1992).

Freeze-Thaw

        Compacted clay is vulnerable to damage from freezing
(Othman et al., 1994). At the levels of compressive
stress expected in landfill covers, hydraulic conductivity
after freeze-thaw is typically 10 to 1,000 times larger
than before freeze-thaw.

        Shan and Daniel (1991) subjected a geotextile-
encased, adhesive-bonded GCL to five freeze/thaw cycles
and found that the hydraulic conductivity did not change.
Similar results have been obtained by commercial testing
laboratories for other GCL products. Hewitt (1992)
performed large-scale tests in tanks, subjecting intact
and overlapped panels of several commercial GCLs to 3
cycles of freeze-thaw. Hewitt found that most of the GCLs
underwent little or no increase in hydraulic conductivity
as a result of freeze-thaw. Available data indicate that
the high shrink-swell capability of bentonite gives
bentonite the ability to self-heal if any alteration
occurs from freeze/thaw cycles. Geosynthetic clay liners
appear to have a much better capacity to remain undamaged
after freeze-thaw than conventional compacted clay liners.

Comparison of Barrier Materials

Many regulatory agencies have traditionally required a low-hydraulic-conductivity, compacted soil liner (or the equivalent) as the primary hydraulic barrier within landfill covers. The thickness of a compacted soil liner typically ranges from 300 to 600 mm, and the maximum allowable hydraulic conductivity is typically 10^{-7} cm/s. Geomembranes have been required by the EPA for hazardous waste landfill covers since the mid 1980s, and for MSW landfills since 1993. However, the MSW rules allow states to approve alternative designs. Geosynthetic clay liners are not required; the issue is whether they ought to be allowed as a substitute for compacted, clayey soil.

Table 3 summarizes the relative advantages and disadvantages of the three barrier materials. Environmental stresses such as freeze-thaw, wet-dry, and distortion caused by differential settlement are much more damaging to low-hydraulic-conductivity compacted soil liners than to geomembranes or GCLs. On the other hand, the interfacial shear strength between a geomembrane or GCL and the adjacent material may limit the steepness of side slopes on which geomembranes and GCLs can be used. Also, thin barrier layers such as geomembranes and GCLs are more vulnerable to construction damage or post-construction puncture, although the consequences of an occasional, unanticipated puncture are much less severe in a final cover system than a bottom liner system.

Cost is sometimes a factor in selecting materials for final cover systems. At the present time, the installed cost for geomembranes and GCLs are both on the order of $5 per square meter ($0.50 per square foot). If suitable material is available locally, the typical cost of a compacted clay liner is about the same as geomembranes and GCLs. If material has to be hauled a large distance, or bentonite blended with soil, the cost is likely to be several times greater. Geomembranes and GCLs can be installed much more quickly, and, because they are thinner than compacted soil, take up less space.

Regulators and design engineers should weigh the pluses and minuses of the alternative materials and then select suitable materials depending upon the specific requirements of individual projects. Too often a low-hydraulic-conductivity, compacted soil barrier is arbitrarily selected when a careful analysis would show that it is not the best material to use.

Table 3. Principal Advantages and Disadvantages of Barrier Materials.

Material	Advantages	Disadvantages
Low-Hydraulic-Conductivity Compacted Soil	1. Long history of use. 2. Regulatory approval is virtually assured. 3. Large thickness ensures that layer will not be breached by puncture. 4. Large thickness provides physical separation between waste and surface environment. 5. Cost is low if material is locally available.	1. Soil can desiccate and crack. 2. Liner must be protected from freezing. 3. Low resistance to cracking from differential settlement. 4. Difficult to compact soil above compressible waste. 5. Suitable soils not always locally available. 6. Difficult to repair if damaged. 7. Slow construction.
Geomembranes	1. Rapid installation. 2. Virtually impermeable to water if properly installed. 3. Low cost. 4. Not vulnerable to desiccation or freeze-thaw damage. 5. Can withstand large tensile strains. 6. Low weight and volume consumed by liner. 7. Easy to repair.	1. Potential strength problems at interfaces with other materials. 2. Geomembranes can be punctured during or after installation.
Geosynthetic Clay Liners	1. Rapid installation. 2. Very low hydraulic conductivity to water if properly installed. 3. Low cost. 4. Excellent resistance to freeze-thaw. 4. Can withstand large differential settlement. 5. Excellent self-healing characteristics. 6. Not dependent on availability of local soils. 7. Low weight and volume consumed by liner. 8. Easy to repair.	1. Low shear strength of hydrated bentonite. 2. GCLs can be punctured during or after installation. 3. Dry bentonite (e.g., at time of installation) is not impermeable to gas. 4. Potential strength problems at interfaces with other materials.

Conclusions

The following conclusions are drawn:

1. Low-hydraulic-conductivity compacted soil liners by themselves are not necessarily the best type of material to use for a barrier layer in landfill covers because the soil is vulnerable to damage from wet-dry cycles, freeze-thaw at some locations, and cracking caused by differential settlement.

2. Geomembranes, if properly installed and protected, are virtually impermeable to water and do not suffer from the problems noted above for compacted soil liners. Interfacial shear with other components of the system may limit the steepness of slopes.

3. Geosynthetic clay liners are a better overall choice than compacted, clayey soil for many final covers because GCLs can better resist wet-dry cycles, freeze-thaw conditions, and differential settlement. However, the low strength of hydrated bentonite, and/or potential low strength of interfaces with a GCL, may limit the steepness of slopes.

No barrier material is without potential problems. All have advantages and disadvantages. For each particular site, materials should be selected that will best perform the required functions.

References

Austin, T. 1992. Landfill-Cover Conflict. *Civil Engineering*. December: 70-71.

Corser, P. and Cranston, M. 1991. "Observations on Long-Term Performance of Composite Clay Liners and Caps," *Proceedings*, Geosynthetics Design and Performance, Vancouver Geotechnical Society, Vancouver, British Columbia.

Hewitt, R.D. (1994), "Hydraulic Conductivity of Geosynthetic Clay Liners Subjected to Freeze/Thaw," M.S. Thesis, Univ. of Texas at Austin, 103 p.

Koerner, R.M. 1994. Designing with Geosynthetics, 3rd Edition, Prentice-Hall, Inglewood Cliffs, New Jersey, 783 p.

LaGatta, M. D. 1992. "Hydraulic Conductivity Tests on Geosynthetic Clay Liners Subjected to Differential Settlement," M.S. Thesis, University of Texas, Austin, Texas.

Leonards, G.A., and J. Narain (1963), "Flexibility of Clay and Cracking of Earth Dams," *Journal of the Soil Mechanics and Foundations Division*, ASCE, Vol. 89, No. SM2, pp. 47-98.

Jessberger, H.L., and Stone, K. 1991. "Subsidence Effects on Clay Barriers," *Geotechnique*, Vol. 41, No. 2, pp./ 185-194.

Melchior, S., Berger, K., Vielhaber, B., and Miehlich, G. 1994. "Multilayered Landfill Covers: Field Data on the Water Balance and Liner Performance," *In-Situ Remediation: Scientific Basis for Current and Future Technologies*, G.W. Gee and N.R. Wing (Eds.), Battelle Press, Columbus, Ohio, pp. 411-425.

Montgomery, R.J., and Parsons, L.J. 1989. "The Omega Hills Final Cap Test Plot Study: Three Year Data Summary," Presented at the 1989 Annual Meeting of the National Solid Waste Management Association, Washington, DC.

Murphy, W.L., and Gilbert, P.A. 1985. "Settlement and Cap Subsidence of Hazardous Waste Landfills," U.S. Environmental Protection Agency, EPA/600/2-85-035, Cincinnati, Ohio.

Othman, M.A., Benson, C.H., Chamberlain, E.J., and Zimmie, T.F. 1994 "Laboratory Testing to Evaluate Changes in Hydraulic Conductivity of Clays Caused by Freeze-Thaw: State-of-the-Art," *Hydraulic Conductivity and Waste Contaminant Transport in Soils*, ASTM STP 1142, D.E. Daniel and S.J. Trautwein (Eds.), American Society for Testing and Materials, Philadelphia, pp. 227-254.

Shan, H.Y., and Daniel, D.E. 1991. "Results of Laboratory Tests on a Geotextile/Bentonite Liner Material," *Geosynthetics 91*, Industrial Fabrics Association International, St. Paul, MN, Vol. 2, pp. 517-535.

Tschebotarioff, G.P., and DePhillippe, A.A. 1953. "The Tensile Strength of Disturbed and Recompacted Soils," *Proceedings*, Third International Conference on Soil Mechanics and Foundation Engineering, Switzerland, Vol. 1, pp. 207-210.

U.S. Environmental Protection Agency. 1989. "Technical Guidance Document, Final Caps on Hazardous Waste Landfills and Surface Impoundments," EPA/530-SW-89-047, Washington, DC.

Alternative Landfill Cover Demonstration

Stephen F. Dwyer[1]

Abstract

The Alternative Landfill Cover Demonstration is a large-scale field test to compare and document the performance of alternative landfill cover technologies of various costs and complexities for interim stabilization and/or final closure of landfills in arid and semi-arid environments. Test plots of traditional designs recommended by the US Environmental Protection Agency (EPA, 1991) for both RCRA Subtitle 'C' and 'D' regulated facilities have been constructed. These will serve as baselines for comparison to alternative covers. The alternative covers designed specifically for dry environments will be constructed in 1996. The covers will be tested under both ambient and stressed conditions. All covers will be instrumented to measure water balance variables and soil temperature. An on-site weather station will record all pertinent climatological data.

A key to acceptance of an alternative environmental technology is seeking regulatory acceptance and eventual permitting. The lack of acceptance by regulatory agencies is a significant barrier to development and implementation of innovative cover technologies. Much of the effort on this demonstration has been toward gaining regulatory and public acceptance. The demonstration is working with regulatory authorities and public interest groups toward the possibility of interstate permitting of alternative landfill cover technologies.

Introduction

The Departments of Energy and Defense have begun a clean-up of their facilities that is expected to cost hundreds of billions of dollars. These cost estimates, however, are based on "state-of-the-art" technologies, of which many are inadequate. Consequently, work has begun on the development or improvement of environmental restoration and management technologies. One particular area being researched is landfill covers. As part of their ongoing environmental restoration activities, the US

---

[1] Principal Investigator, Sandia National Laboratories, Albuquerque, NM 87185-0719.

Department of Energy (DOE) and the US Department of Defense (DOD) have thousands of radioactive, hazardous, mixed waste, and sanitary landfills to be closed in the near future (Hakonson et al., 1994). These sites, as well as mine and mill tailings piles and surface impoundments, all require either remediation to a 'clean site' status or capping with an engineered cover upon closure. Additionally, engineered covers are being considered as an interim measure to be placed on contaminated sites until they can be remediated. The Alternative Landfill Cover Demonstration (ALCD) is a large-scale field test taking place at Sandia National Laboratories, located on Kirtland Air Force Base in Albuquerque, New Mexico. It will compare and document the performance of alternative landfill cover technologies of various costs and complexities for interim stabilization and/or final closure of landfills in arid and semi-arid environments.

Health and environmental risk assessment work is currently underway at the various federal facilities. These studies will almost certainly recommend capping technologies to be widely used rather than the remediation of landfill sites to a 'clean site' status. Early estimates to excavate buried hazardous waste and completely remediate the site to a clean position are about 20 times more expensive than capping it with an engineered cover in place. (Hakonson et al., 1994).

Traditional Cover Designs

The ALCD will install traditional designs as baseline covers to compare and contrast alternative covers against. The ALCD is comprised of phased construction due to funding limitations. Phase I constructed in 1995 consists of three covers.

1) a traditional Compacted Clay Cover meeting minimum requirements of Subtitle 'C' with the following profile - a 60 cm vegetation soil layer over a 30-cm sand drainage layer over a geomembrane over the 60 cm compacted clay barrier layer (EPA, "Construction....", 1994).
2) A basic Soil Cover that meets minimum requirements set forth in Subtitle 'D' - comprised of a 15 cm vegetation soil layer over an 45 cm compacted soil barrier layer (DOE, 1993).
3) A Geosynthetic Clay Cover identical to the traditional compacted clay cover with the exception that the problematic clay barrier layer will be replaced with a manufactured sheet installed in its place - a geosynthetic clay liner (GCL).

Phase II designs will be finalized in 1995 and are scheduled to be constructed in the summer of 1996. These designs include a capillary barrier, dry barrier, anisotropic barrier, and enhanced evapotranspiration cover.

Traditional covers presently in use for Subtitle 'C' and 'D' regulated facilities as recommended by the US Environmental Protection Agency (EPA) are used throughout the US with little regard for regional environments. Experience in the

Western United States has shown these designs to be vulnerable to such things as desiccation cracking when installed in arid environments. An EPA design guideline states "In arid regions, a barrier layer composed of clay (natural soil) and a geomembrane is not very effective. Since the soil is compacted 'wet of optimum', the layer will dry and crack." (EPA, 1991) The clay barrier layer in the traditional Subtitle 'C' cover must be constructed so as to yield a maximum hydraulic conductivity of 1×10^{-7} cm/sec. To achieve this, the soil often requires amendment (e.g. mixed with bentonite) and should be compacted 'wet of optimum'. Compacting this layer 'wet of optimum' in dry environments leads to drying and cracking of this layer. This desiccation cracking can cause serious problems with its ability to block the infiltration of water. The basic soil cover used with Subtitle 'D' covers has a barrier layer that is also subject to desiccation cracking, as well as, deterioration due to freeze/thaw cycles among other problems. These traditional covers such as the Compacted Clay Cover not only have inherent problems but are very expensive and difficult to construct.

A study by the EPA (EPA, 1988) of existing landfill performance revealed that many have serious problems, from groundwater contamination to serious ecological impacts such as killing flora and fauna. Virtually all parts of the nation have experienced water contamination due to leachate leaking from landfills in some degree (EPA, 1988). Not all of these problems are the result of inadequate covers. Many older landfills were crudely installed (e.g., poor siting, inadequate or lack of liner) thus destined for failure, but these problems can be mitigated by capping the entire landfill with a good quality final cover.

Alternative Acceptance

Without regulatory acceptance, promising technologies will die with their inventors. The ALCD has been committed from the start to get regulatory and public acceptance of the project and the technologies presented in the demonstration.

A study performed by the University of North Dakota (Wentz, 1989) concluded that the deciding factors affecting which hazardous waste management technology is to be used are from most important to least important: 1) Government Regulations, 2) Economics, 3) Public Relations, and 4) Process / Technology

Because permits can be difficult to get and their has been only minimal work done promoting alternative covers based on regional environmental requirements, many design engineers are reluctant to stray far from conventional designs.

A commercialization plan written for the ALCD concluded that for the technologies in the project to be accepted, the demonstration must meet several criteria (Dwyer, 1994). The covers must be readily comparable. The data must be real and obtained fairly. Key individuals and groups, independent to the project design, serve as reviewers to ensure the criteria is met. The Phase I designs were sent out for review first to a group of technical peers that were independent of the project and deemed industry experts. This review helped ensure the technical validity of the designs to be constructed in Phase I. Comments were gathered from the reviewers and included in the designs.

This revised test plan was then sent to regulatory representatives from Environmental Departments from many of the western states including New Mexico, Arizona, California, Nebraska, Nevada, South Dakota, Texas, and Utah. It was also sent to representatives from the EPA Regions VI and VIII offices. Comments from this review were also incorporated into the design package.

Politicians and thus regulators are becoming more sensitive to special interest groups concerns and are therefore encouraging participation with these groups when permitting projects. The ALCD has received endorsement by a committee from a western states' and federal government initiative to accelerate and improve clean up of federal lands. This initiative originated in 1992, when the Western Governors Association, the Secretaries of Defense, Energy, and Interior, and the Administration of the Environmental Protection Agency formed a federal advisory committee to cooperate on the cleanup of federal waste management sites in the region. This committee, known as the Committee to Develop On-Site Innovative Technologies (DOIT Committee), has sought the guidance of key players to help identify, test and evaluate more cooperative approaches to deploying promising innovative waste remediation and management technologies in order to clean up federal waste sites in an expeditious and cost-effective manner.

The DOIT Committee's primary goal with regard to the ALCD is to assist with the eventual acceptance of new technologies that come from the demonstration and inclusion of landfill permitting in an inter-state reciprocity program the Committee is attempting to finalize.

Yet, another review process included sending a general overview of the demonstration to members of the DOIT Committee and special interest groups identified by the DOIT Committee. These interest groups included representatives from such entities as environmental activist groups like the Sierra Club, Indian tribes, government agencies, neighborhood associations, local businesses, engineering firms, and politicians. Over 1000 groups received a package. Comments were forwarded through the Western Governors Association for consideration. The majority of these comments centered on questions rather than comments and on praise for getting them involved early in the process. Much interest was invoked as a result of this. Periodic meetings were held with these representatives of some of the special interest groups, Western Governor's Association, regulatory agencies (predominantly from New Mexico), New Mexico State Legislature, and Sandia National Laboratories. These meetings kept interested parties apprised of advancements, progress, and answered questions and concerns. These meetings continue on about a bimonthly basis.

Demonstration Description

The ALCD is a series of large-scale landfill test covers constructed side-by side for comparison. (see figure 1). Future test cover construction will continue this side-by-side arrangement. The various covers will be compared based on their performance, cost, and ease of construction.

The ALCD is not intended to showcase any one particular cover system. It is intended to compare and contrast different cover systems in a dry environment. Information gained from the demonstration can then be used by others when choosing between cover designs or when applying for the permitting of one of the cover systems.

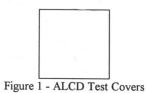

Figure 1 - ALCD Test Covers

The covers are each 13 m wide by 100 m long. The 100 m dimension was chosen because it is fairly representative of hazardous and mixed waste landfills found throughout the DOE complex (approximately 2 acres in surface area).

General site preparation included bringing utilities (water and power) to the site for the instrumentation and stress testing. The site was cleared and grubbed after which the top 15 cm of topsoil was excavated and stockpiled. This topsoil was reused as the top 15 cm of the covers. The covers were designed so that the site cut excavations were approximately equal to the soil requirements for the covers. The subgrade below each cover was compacted to 95% of maximum dry density (ASTM D698). The only soil hauled in from off-site used in the construction of the covers was the bentonite added to the barrier layer in the Compacted Clay Cover and the sand for the drainage layers. All covers were constructed with a 5% slope in all layers.

Phase I Cover Descriptions

The Compacted Clay Cover installed was designed to meet minimum requirements from Subtitle 'C' regulated landfills (EPA, 1991). It is 1.5-meters thick. The typical profile for this cover consists of three layers (see figure 2). The bottom layer is a 60 cm thick barrier layer. The barrier layer's primary purpose is to prevent the downward movement of water into underlying waste. It was constructed of native soil mixed with 6% bentonite. The bentonite was required because the native soil used did not have an adequate amount of fines in it. The lifts were installed in maximum 22 cm loose lift thicknesses and compacted to 98% of maximum dry density (ASTM D698). The soil during fill and compaction was kept at a water content two to four percent 'wet of optimum'. The soil was compacted 'wet of optimum' so as to remold the soil. The combination of the compaction requirements, soil amendment, and placement 'wet of optimum' was done to yield a maximum hydraulic conductivity of 1×10^{-7} cm/sec. Laboratory soil tests revealed that the amended soil compacted to these requirements had a hydraulic conductivity lower than 1×10^{-8} cm/sec. A lesser percent of bentonite would have been acceptable based on laboratory tests but it was decided, based on past experience and discussions with colleagues, that 6% bentonite is probably the minimum practical amount that one can add to get a fairly uniform application and mix. Partial penetrating kneading compaction was used rather than full penetrating so as not to damage instrumentation (TDR probes, thermocouples and associated cable) placed on or in the preceding lift. Each lift was scarified an inch or two prior to placement of the next lift. This was done in an effort to eliminate the interface between lifts. An interface left could provide a pathway for water to run along until it found a crack in the underlying lift. Organic material and stones larger than 2 cm in diameter were removed from soil prior to placement. This was done in order to remain consistent with the soil preparation for the hydraulic conductivity laboratory tests. The top lift of the barrier layer was compacted smooth. Objects larger than 13 mm were removed from the surface or near the surface so as not to damage the geomembrane that was placed on top of this barrier layer.

A 40 mil linear low density polyethylene geomembrane was placed directly on top of the clay barrier layer. The surface of the clay barrier layer was smooth-roll compacted and prepared to allow for intimate contact between it and the under-surface of the geomembrane to essentially get a composite barrier layer. Continuous contact was emphasized in an effort to eliminate gaps between the geomembrane and soil barrier layer where any water that has passed through the geomembrane would have a pathway to run and find a crack in the underlying soil layer.

The middle layer is a 30 cm thick drainage layer. The primary purpose of the drainage layer is to quickly route any water that has passed through the vegetation layer laterally to collection drains normally located at the perimeter of the landfill. This layer was constructed of sand placed directly on the geomembrane. It was not compacted to a level beyond that received as a result of ordinary construction activities. The average hydraulic conductivity of the sand installed was 1×10^{-1} cm/sec which is an order of magnitude better than the minimum 1×10^{-2} cm/sec called for in Subtitle 'C'. A nonwoven polyester needlepunched geotextile was placed directly on top of the sand drainage layer. This served as a filter between the drainage layer material and top layer.

The top layer is a 60 cm thick vegetation layer. This layer's primary purpose is to provide for vegetation growth, erosion protection, and protect the underlying layers from such things as harmful freeze/thaw cycles. It allows for storage of infiltrated water that can later be evaporated. It is 45 cm of native soil covered by 15 cm of topsoil. This entire layer was placed and left uncompacted. Stones larger than 50 mm in diameter were removed from this soil prior to replacement.

After completion of all covers, the surfaces were seeded with native grasses. A cereal straw mulch was spread over the seed with a tackifier applied to help prevent the wind from blowing it away. The seed and fertilizer mixture applied was chosen with the assistance of the New Mexico State Highway Department and is consistent with vegetation found in the immediate area. The vegetation provides for erosion control and transpiration of water through the plants from the vegetation layer.

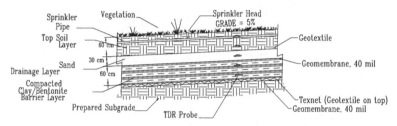

Figure 2 - Compacted Clay Cover typical section.

The basic Soil Cover installed met minimum requirements for Subtitle 'D' governed landfills (DOE, 1993). This cover is 60 cm thick (see figure 3). It is constructed of essentially two layers. The bottom layer is a 45 cm thick compacted soil barrier layer. Only native soil was used in this layer. Organics and objects larger than 50 mm in diameter were removed prior to placement. This barrier layer was constructed by placing loose lifts no thicker than 22 cm. The soil was compacted to 95% of maximum dry density (ASTM D698). Partial rather than full penetrating kneading compaction was used so as not to damage underlying instrumentation. The soil was placed as specified to meet the maximum 1×10^{-5} cm/sec requirement of Subtitle 'D' regulated facilities. Laboratory test results yielded saturated hydraulic conductivity results on the barrier layer soil between 5.5×10^{-6} cm/sec and 5.1×10^{-7} cm/sec.

The top vegetation layer is 15 cm of topsoil loosely laid. This layer provides for vegetation growth and erosion protection.

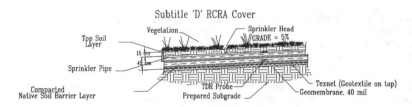

Figure 3 - Soil Cover typical section.

The third cover is called a Geosynthetic Clay Liner (GCL) cover (see figure 4). It is identical to the Compacted Clay Cover installed with one exception. The

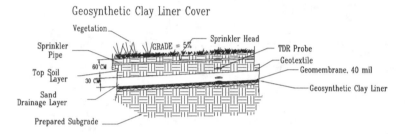

Figure 4 - GCL Cover typical section.

compacted clay barrier layer was replaced by a single manufactured sheet known as a GCL. Therefore the GCL is the bottom layer covered with a geomembrane, drainage layer and vegetation layer, respectively. The GCL sheet installed is a composite of two nonwoven geotextile fabrics sandwiching a layer of bentonite. The hydraulic conductivity of the GCL is 5×10^{-9} cm/sec.

Apart from the three covers installed, two smaller test pads were also constructed adjacent to the test pads. The test pads are 7.3 m by 7.3 m. One pad is a Subtitle 'C' compacted clay barrier layer with the other being a Subtitle 'D' compacted soil barrier layer. They were constructed exactly like the barrier layers in the full size test plots. A double-ring infiltrometer was installed on each test pad to measure in-situ hydraulic conductivity (ASTM D5093). These infiltrometers will be left in-place to measure hydraulic conductivity throughout the duration of the experiment.

Phase I Construction Quality Assurance

A detailed Quality Assurance (QA) Plan was prepared for this demonstration. It adheres closely to that recommended by the EPA (EPA, 1993). The major purpose of this QA process was to provide documentation for those individuals who were unable to observe the entire construction process (e.g., representatives of regulatory agencies, etc.) so that these individuals can make informed judgments about the quality of construction of the ALCD. QA procedures and results were thoroughly documented.

Daily Inspection Reports were prepared that included information about work that was accomplished, tests and observations that were made, and descriptions of the adequacy of the work that was performed. Daily Summary Reports provided a chronological framework for identifying and recording all other reports and aided in tracking what was done and by whom. Inspection and Testing Reports noted field observations, results of field tests, and results of laboratory tests performed on- or off-site. These observations took the form of notes, charts, sketches, or photographs, or a combination of these. Problem Identification and Corrective Measures Reports identified and recorded fixes of problems with material or workmanship that did not meet the requirements of the plans, specifications, or QA Plan. Drawings of Record ('as-built' drawings) were prepared and continually updated to document actual field installations. Final Documentation and Certification took the form of a final report that included all of the aforementioned.

Key meetings were essential to the successful construction of the ALCD. These meetings included a pre-bid meeting held prior to bidding of the contract, a pre-construction meeting held in conjunction with a resolution meeting after the contract was awarded but prior to the start of actual construction activities.

The pre bid meeting was used to discuss the QA Plan and resolve differences of opinion before the project was let for bidding. It also gave the bidders a chance to ask questions and problems which were therefore rectified early on. The resolution / pre-construction meeting allowed for lines of communication, review of construction plans and specifications, emphasize the critical aspects of a project necessary to ensure proper quality, begin planning and coordination of tasks, and anticipate any problems that might cause difficulties or delays in construction. It also allowed for the review of the QA Plan, to make sure that the responsibility and authority of each individual was clearly understood, where procedures to resolve construction problems were established. Periodic progress meetings were held at the job site. These meetings were helpful in maintaining lines of communication, resolving problems, identifying action items, and improving overall quality management.

Materials Quality Assurance was of utmost importance. Materials and their installation were tested to ensure compliance with the design and recorded throughout the construction process. These items included such things as plasticity index, sieve analysis, maximum size stone or debris, placement and compaction, moisture content, bond between lifts, in-situ hydraulic conductivity (ongoing), bentonite content, water content, lift thicknesses, etc.. Geosynthetic installations were also inspected and recorded.

Performance Monitoring and Instrumentation

Objectives to be achieved through monitoring are to: (1) collect continuous water balance data from each cover design and ancillary supporting data for a minimum five-year post-construction period, (2) periodically measure characteristics of the vegetation cover on each plot, and (3) manage the automated computer database.

The test covers were built with half of their length (150 m) sloping at 5% toward the east with the other half sloping toward the west (see figure 1). The western slope will be monitored under ambient conditions (passive monitoring). A sprinkler system was installed in the eastern slope of each cover. This will allow for stress testing of the covers (active monitoring).

Passive monitoring for the first few months will consist of daily on-site observations that will be required to validate system performance and to correct problems. As the bugs in the system are worked out, on-site observations will be necessary less frequently. Continuous data will be obtained for soil moisture status, percolation and interflow, runoff and erosion, precipitation, wind speed and direction, relative humidity, air and soil temperature. Periodic measurements will be obtained on vegetation cover, biomass, leaf area index, and species composition.

Active monitoring will include supplemental precipitation added to hydrologically stress the various barriers systems. Water will be applied using the

ALCD CONTROL & MONITORING INSTRUMENTATION

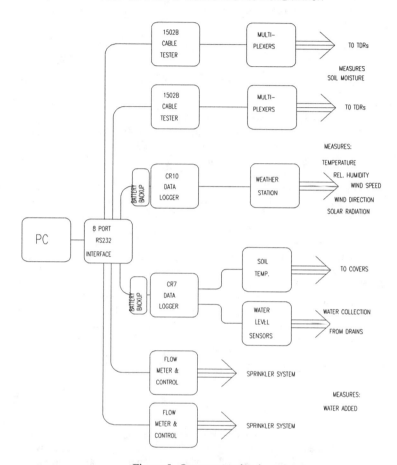

Figure 5 - Instrumentation layout.

sprinkler system tested for rate and uniformity of application. All water will be distributed through electronically controlled flowmeters where quantities discharged will be controlled and measured. Worst case precipitation events such as a 10-, 50-, or 100-year storm can be applied. All other measurements under this precipitation regime would be the same as described above for passive monitoring.

The success of the ALCD will depend heavily on the quality of the monitoring systems and the care taken in their installation and follow-on maintenance. The monitoring equipment was designed for measuring the components of water balance and additional ancillary variables. All monitoring equipment was also designed with a backup for each.

The water balance equation to be used is:

$$E = P - I - R - D - \Delta S; \text{(Equation 2)}$$

where, precipitation (P), surface runoff (R), lateral drainage (D), evapotranspiration (E), soil water storage (S), and percolation or infiltration (I) are the six water balance variables. With the exception of 'E', estimates of all terms in Equation 2 will be obtained with the monitoring systems. Evapotranspiration will be estimated by solving Equation 2 for 'E'. Other measurement variables include erosion, precipitation, relative humidity, barometric pressure, soil temperature with depth, vegetation biomass and cover, and wind speed and direction. Most of the physical attributes will be measured with automated monitoring systems to provide continuous data (see figure 5).

Soil Moisture: Time Domain Reflectometry (TDR) and an associated data acquisition system will be used to provide a continuous record of soil moisture status at various plan locations and depths in each cover profile. The soil moisture will be measured using TDR. PVC pipes were installed strategically in the covers to be used as ports to allow for the use frequency domain reflectometry as a backup. The process of sending pulses and observing the reflected waveform is called TDR. A waveform traveling down a coaxial cable or waveguide is influenced by the type of material surrounding the conductors. If the dielectric constant of the material is high, the signal propagates slower. Because the dielectric constant of water is much higher than most materials, a signal within a wet or moist medium propagates slower than in the same medium when dry. Ionic conductivity affects the amplitude of the signal but not the propagation time. Thus, moisture content can be determined by measuring the propagation over a fixed length probe embedded in the medium being measured.

Soils with a high water content lengthen the propagation time and this is reflected as an apparent increase in the distance traveled by the pulse. Soils with a high water content and a high electrical conductivity rapidly attenuate the voltage pulse before it is reflected back to the source. If the attenuation is great enough there is no return signal and the probe cannot be used. This is essentially what happened when commercially available probes were tested in soils representative of the barrier layer in the Compacted Clay Cover -- no useful signal was reflected back. The high density, water content and sodium bentonite addition made TDR a challenge to use. Because of the problems with the commercially available TDR systems, a system was designed to overcome these problems. The design began with perhaps the single most important element and weak link - the probe. Such things as: rod length, rod spacing, number of rods (2 or 3), coating the rods (with several different types of coatings, coating thicknesses, and surface preparation), total coax cable lengths, rod diameter, low loss coaxial cable, and inserting diodes were experimented with to perfect the probes.

The final design yielded excellent waveforms under the most trying of circumstances. 256 probes were fabricated, installed in the three covers, and multiplexed (SDMX50 by Campbell Scientific) back to a set of Cable Testers (Tektronix 1502B). After fabrication of the probes but prior to their installation, each probe was individually calibrated. This calibration process was extremely lengthy and time consuming. Each probe as it would be assembled in the field was inserted in representative soil under a range of moisture contents to develop an algorithm that would yield an accurate moisture content. The TDR system was calibrated to measure the soil moisture content to within +/- 1% of actual moisture content.

Soil Temperature: Thermocouples placed strategically throughout each cover will measure the soil temperature. This data will be used to assist with evapotranspiration studies and for monitoring frost penetration and its affect on the hydraulic conductivity of the soils.

Runoff and Erosion: Runoff and erosion will be measured on an event basis. Surface runoff water will be collected with a gutter system located at the bottom of each slope of each cover. The collected water will be routed to instrumentation that quantifies it. All instrumentation is set up so as to have redundancy in case of a failure in the primary measurement. A data acquisition system is linked to the instrumentation to automatically record and store data. Sediment will be separated from runoff in a settling tank located downstream from the runoff measuring system to provide total soil loss for each runoff event.

Percolation and Interflow: Subsurface flows will be measured. Lateral drainage from each drainage layer will be collected using underdrain systems placed at the bottom of each slope of each cover. The water will be routed to instrumentation that quantifies it. The instrumentation is linked to a data acquisition system to continuously record flow events. Percolation through the barrier layer for each cover will be collected using a geomembrane under a geonet that routes the water to an underdrain collection system. Both percolation and interflow will be routed via drains to the flow monitoring system. Measurement redundancy is built into the system to reduce the chances of losing data due to equipment failure or power loss and to verify correctness of results obtained. To avoid problems with inclement weather, all monitoring instrumentation is housed in a shelter.

Meteorology: A complete weather station was installed at the ALCD site. Precipitation, air temperature, wind speed and direction, relative humidity, and solar radiation will be continuously recorded. The meteorological measurements will be made with automated equipment coupled to the data acquisition system.

Vegetation: Attributes of the vegetation will be measured seasonally throughout the study to relate to changes in erosion and evapotranspiration. A point frame will be used to evaluate cover and leaf area. Biomass will be determined by clipping and weighing oven-dried samples collected from subplots within each cover design. Species composition will be measured using line transacts and/or quadrants. This activity commences upon completion of each respective test plot.

Phase II Designs

(1) A Capillary Barrier primarily comprised of a fine-grained layer of soil placed over a coarse-grained layer. The design emphasizes a sufficient contrast between the hydraulic conductivities of the fine-grained layer versus the coarse-grained layer. This contrast lends to the effect that flow through the cover is greatly slowed under unsaturated conditions. Also, under unsaturated conditions, the hydraulic conductivity of the coarse-grained soil is significantly lower than that of the fine-grained soil layer. If saturation is reached however, the cover fails.

(2) An Anisotropic Barrier that attempts to limit downward movement of water while encouraging lateral movement of water. This cover is composed of a layering of capillary barriers. The various layers are enhanced by varying soil properties and compaction techniques that lead to the anisotropic properties of the cover.

(3) A Dry Barrier which is an enhanced capillary barrier. It is comprised of a fine-grained soil over a coarse-grained material. The preferred coarse-grained material is volcanic tuff as found in the Los Alamos, NM area or scoria. This material lends to water storage in the coarse-grained layer. The design assumes that the first layer will store water and allowed to evapotranspirate. A capillary barrier exists with the contrast in hydraulic conductivity between the two materials. Any water entering the coarse layer where it is stored in the material will be dried by air that enters the coarse layer through supply vents strategically located through the cover (goose-necked pipes). The air then exits through passive air vents such as a common roof turbine secured to a return pipe that extends from the coarse layer and are strategically located throughout the cover.

(4) An enhanced evapotranspiration (ET) Cover that is a soil cover with an engineered vegetative covering. This cover encourages water storage and enhanced ET year-round rather than just during the growing seasons.

Expected Benefits of the ALCD

The demonstration is expected to provide performance and cost data for cover components and systems that are more applicable to Western climatic conditions. A direct comparison between conventional and alternative designs will be available. The "active" testing will permit data to be collected under extreme and accelerated conditions. This information will allow those responsible for the development of landfill cover design guidance to have a defensible basis for the transition from designs suited for the eastern US to those more suited to the western US.

This demonstration project should provide valuable field data for validation of the HELP Modelv3 (EPA, "HELP...., 1994). The demonstrations will provide data from extreme conditions (arid and high-precipitation) that will be used to test the capabilities of the HELP Modelv3. This computer program by the EPA is routinely used by practicing engineers to assist with design and application for permits.

The probable outcome of this demonstration is the acceptance of alternative cover designs that are significantly less costly than conventional designs. Given the thousands of acres of buried waste sites to be covered, the payoff from this demonstration may be on the order of many millions of dollars in savings.

Future papers will discuss the comparison of construction specifics and costs between the various covers. Also, results comparing and contrasting the water balance variables between the covers as well as how well field results compared with predictive models such as the Hydrologic Evaluation of Landfill Performance (HELP).

Acknowledgments

The work was supported by the United States Department of Energy Office of Technology Development through the Landfill Stabilization Focus Area Low Level and Other Contaminants in Arid Environments product line. G. Reynolds, J. Lopez, T. Steinfort, and S. Bayliss were of invaluable service in assisting with the design of instrumentation, field QA work, and field engineering.

REFERENCES:

1.   DOE (1993). "Closure of Municipal Solid Waste Landfills (MSWLFs). Office of Environmental Guidance, Document: EH-231-036/0793.

2.   Dwyer, S.F. (1994). *Commercialization Plan: Alternative Landfill Cover Demonstration.* Department of Energy, Office of Technology Development.

3.   EPA, Office of Solid Waste (1988). "Criteria for Municipal Solid Waste Landfills" (40CFR Part 258) Subtitle D of Resource Conservation and Recovery Act (RCRA). Draft - Background Document. Case Studies on Ground-Water and Surface Water Contamination from Municipal Solid Waste Landfills: EPA/530-SW-88-040. PB88-242466.

4.   EPA (1991). "Design and Construction of RCRA / CERCLA Final Covers." Seminar Publication: EPA/625/4-91/025, May 1991.

5. EPA (1993). "Quality Assurance and Quality Control for Waste Containment Facilities." Technical Guidance Document. EPA/600/R-93/182.

6. EPA (1994). "Construction Quality Assurance/Construction Quality Control (CQA/CQC) for Waste Containment Facilities; and Hydrologic Evaluation of Landfill Performance (HELP) Model." Seminar Publication EPA/625/K-94/001.

7. EPA (1994). "The Hydrologic Evaluation of Landfill Performance (HELP) Model, Engineering Documentation for Version 3." EPA/600/R-94/168a.

8. Hakonson, T.E., Bostick, K.V., Trujillo, G., Manies, K.L., Warren, R.W., Lane, L. J., Kent, J.S., Wilson, W. (1994). *Hydrologic Evaluation of Four Landfill Cover Designs at Hill Air Force Base, Utah.* Los Alamos National Laboratories document: LA-UR-93-4469.

9. Wentz, Charles A (1989). *Hazardous Waste Management.* McGraw-Hill Publishing Company, NY, NY 10020.

Innovative Cover Design for 2:1 Slopes Complying With California Closure Criteria

by
Jeffrey G. Dobrowolski, P.E.[1]
A.S. Dellinger, P.E.; Member, ASCE[2]

Background

Closure of inactive landfills, while preventing harmful effects to the environment, is a challenge facing all landfill engineers. The need to balance closure costs with the benefits associated with environmental mitigation must be considered when designing landfill final covers.

California is recognized as having some of the most stringent environmental standards in the United States. These standards, coupled with seismic stability requirements, necessitate creative and innovative landfill cover designs.

The City of Los Angeles is responsible for developing final closure plans for the Toyon Canyon Sanitary Landfill, a 0.36 square kilometer (90 acre) canyon fill site with refuse depths of 88 m (290 feet) on average. The front face (sloped area) of Toyon Canyon Landfill encompasses 0.16 square kilometers (40 acres), with 10 benches or terraces; bench heights range from 9 to 12 m (30 to 40 feet). This site ceased waste disposal in 1985, with disposal of 14.5 million metric tons (16 million tons) of refuse, but has yet to implement a final cover system. The challenge in designing a final cover system for this landfill was placement of cover material which satisfied permeability limits while meeting seismic slope stability needs. What makes this site unique is the presence of slopes at 2:1 (2h:1v); current manufacturers of cover products cannot satisfy seismic stability factors of safety (FS) for placement on slopes with angles exceeding 26 degrees.

[1]Engineer, City of Los Angeles, Bureau of Sanitation, 419 S. Spring Street, Suite 800, Los Angeles, CA 90013, (213) 893-8210.

[2]Ibid.

Purpose
This paper discusses the near, mid, and far-field seismic criteria necessary for developing slope factors of safety when designing final cover systems. These seismic criteria will be related to the Toyon Canyon Sanitary Landfill, particularly their impact and necessary considerations for 2:1 slopes.

After demonstrating the techniques for determining seismic stability needs, this paper will address the development of performance standards for final cover material. Specifically, it will show the process the City of L.A. undertook in conducting laboratory modeling and testing of proposed final cover material supplied by manufacturers and the development of performance specifications used in purchasing the final cover. Finally, the relationship of the final cover system to other landfill systems, including leachate collection and landfill gas (LFG) collection, will be discussed.

The greatest challenge to the final cover design comes with evaluating the structural integrity, under both static and pseudo-static (or seismic) conditions.

California Closure Requirements
Landfill closure is generally administered by the California State Integrated Waste Management Board (CIWMB). However, the Toyon Landfill is unique in that the lead regulatory agency for closure design is the California Regional Water Quality Control Board (CRWQCB) - Los Angeles Region. The CRWQCB has retained its lead role throughout closure because they regulated waste discharge requirements when the landfill was operating.

As a Class III landfill, the site is subject to the California Code of Regulations (CCR), Title 14, Division 7, Chapter 3, Article 7.8, Chapter 5, Article 3.4, and Title 23, Division 3, Chapter 15, Articles 4,5,8, and 9. The site is also governed by Title 40, Part 258 of the Code of Federal Regulations. The following table presents a detailed list of the issues required under final closure:

California Regulatory Requirements for Class III Landfill Closure		
Closure and Postclosure Performance Standards	Final Closure and Postclosure Plans	Final Cover Design
Final Grading Design	Slope Performance Evaluation	Final Drainage Control
Slope Protection and Erosion	Leachate Monitoring and Control	Landfill Gas Monitoring and Control
Final Site Security	Closure Construction Quality Assurance	Postclosure Land Use
Postclosure Leachate Monitoring Program	Postclosure Water Quality Monitoring Program	Postclosure Landfill Gas Monitoring Program
Certificate of Closure	Postclosure Inspection and Maintenance Program	

Only the following areas will be discussed in this paper: cover design, slope stability in a seismic zone, leachate control and monitoring, and landfill gas monitoring and control.

Cover Design

The minimum criteria for landfill closure design in California is governed under CCR Titles 14 and 23. Specifically, these regulations require a cover system comprised of the following parts: a minimum 0.6 m (2 feet) thick foundation layer, a hydraulic barrier layer with permeability less than or equal to 1×10^{-6} cm/sec, a minimum 0.3 m (1 foot) thick vegetative soil cover, and a minimum grade over the landfill surface of 3%. California regulations are more stringent than that promulgated under Federal law, and, therefore, govern.

Since 1985, when refuse disposal ceased, cover material has been added to the slopes and top deck to maintain minimum drainage. Geotechnical investigations indicate a varying thickness of cover material throughout this site ranging from 6 m (20 feet) on the top deck to 0.3 m (1 foot) on the slopes. Recognizing that the cover thickness varied on the slopes and presented a health and environmental hazard from refuse exposure, the Bureau decided to cut existing cover material from the top deck and fill on the slopes. In effect, the slopes were treated as if no cover material was in place and that construction of this new 0.6 m (2 feet) thick foundation layer would satisfy the required minimum thickness. Approximately 229,000 m³ (300,000 cubic yards) of material will be imported to facilitate foundation layer construction.

Importation of clay to function as the hydraulic barrier layer was deemed unacceptable as the site was located within a public park, and importation of material satisfying the maximum permeability of 1×10^{-6} cm/sec, 0.3 m (1 foot) thick, would require an additional 115,000 m³ (150,000 cubic yards) of material, or 100 truck trips each day for 120 days.

The Bureau elected to utilize a geosynthetic clay liner (GCL) as a system which would satisfy permeability limits. This liner could be hauled to the site and easily placed over the top deck. Several manufacturers are available which meet these criteria. However, placement of a GCL system on the slopes presented a challenge in meeting minimum seismic stability requirements as detailed below.

The Toyon Landfill is underlain with natural springs and leachate seeps have been observed emanating from the landfill slopes after periods of heavy rains. The presence of leachate also provided a buoyancy effect on the cover system, further reducing slope stability FS. To aid in the removal of this leachate, a high-density polyethylene (HDPE) geonet was provided. This geonet is located below the hydraulic barrier and within the foundation layer.

Regulations mandated a 0.3 m (1 foot) minimum vegetation layer. In order to preserve the integrity of the cover system, specifically the hydraulic barrier layer, the vegetation layer should be as thick as possible. However, seismic analysis indicated that slope stability FS decreased as the vegetative layer thickness was increased. Therefore, the vegetative layer was kept at the 0.3 m (1 foot) minimum thickness, and the landscaping/irrigation systems were modified accordingly.

Static Slope Stability

Static slope stability analyses were performed using two- and three-dimensional limit equilibrium methods. Two-dimensional analyses were performed for two groundwater surfaces: a deep groundwater profile based on water levels measured in the existing off-site monitoring wells, and a shallow surface inferred from historical leachate seeps on the front face (Figure 1). Analyses results indicate that, for the "average" representative strength values of the refuse, the calculated static FS ranges from 1.3 to 1.6 for low groundwater and 0.7 to 0.9 for high groundwater conditions.

Three-dimensional static analysis for the high groundwater condition indicated that the calculated static FS, incorporating three-dimensional effects of the narrow canyon walls, ranges from 1.2 to 1.3; a minimum factor of safety of 1.5 is required for static conditions pursuant to 14 CCR 17777.

The assumed high groundwater surface in the landfill is predominantly due to infiltration of rain through the existing cover, which currently does not provide an impermeable hydraulic barrier. The proposed final cover system will drain perched water in the landfill to a significantly lower level and increase the long-term stability of the landfill.

The long-term static FS, with the final cover installed, was estimated from the results of two-dimensional analysis for the lower groundwater surface (FS equal to 1.3 to 1.6) and is anticipated to increase by 50 percent when including three-dimensional effects. Based on this, the long-term static FS for the closed landfill is estimated to be greater than 1.5.

Seismic Slope Stability

Because the FS calculated for the landfill slope under static conditions is only 1.5, it is clear that a 1.5 FS would not be achieved for the seismic conditions, as required by 14 CCR 17777. Therefore, the analysis of seismic slope behavior focused on evaluating the probable range of seismically induced slope deformation due to maximum credible earthquakes established for the landfill site.

The analysis involves the following steps:

- Selecting representative earthquake time histories from near-field, mid-range, and far-field seismic activities for input in the seismic analysis.

- Determining the seismic responses in the potential sliding masses.

- Calculating the permanent displacement based on a double integration of the acceleration response time history exceeding specified yield acceleration limits.

- Comparing the calculated permanent displacement values to anticipated yield accelerations calculated from pseudostatic analysis.

Following these procedures, the yield acceleration for the landfill was determined to be about 0.07g. Yield acceleration is defined as the average acceleration that produces a horizontal inertia force on a potential sliding mass with a FS of 1.0. The analyses indicated that the largest slope movement may be associated with the maximum far-field credible earthquake, specifically an earthquake of magnitude 8 or greater (on the Richter Scale; M = 8) along the San Andreas Fault about 35.4 km (22 miles) from the landfill. Anticipated permanent slope movement on the order of 0.6 m (2 feet) may occur from this earthquake. The landfill's overall integrity is not likely to be impacted from slope displacement of this magnitude as the landfill contains no rigid structures such as concrete pipes or columns.

Several design alternatives for the final closure cover system were investigated. These alternatives include the use of various GCLs (Claymax, Gundseal , Bentomat , etc.) as the hydraulic barrier. Each was evaluated with respect to the proposed final grade slope stability. The slope stability analyses evaluated the FS for sliding between various interfaces to identify potential surfaces along which the cover could fail. The interface-strength parameters used in the analyses were based on the information provided by the GCL manufacturers, literature, and data obtained from interface shear tests conducted using an apparatus specifically designed to simulate the low shearing stress at the interface.

The analyses indicate that the most vulnerable potential failure surface of the proposed cover system is within the interface of saturated vegetative cover soils and the GCL. The calculated FS under static condition ranges from 1.2 to 1.8 depending on the GCL. Interface shear tests performed under low overburden condition, found that a GCL sandwiched between two sides of nonwoven geotextile (in addition to the woven geotextile sandwiched inside) can provide slope stability meeting regulatory requirements.

In addition to using a high strength GCL, options such as additional mid-slope GCL anchor trenches can also be used to increase slope stability. The mid-slope anchor trenches would reduce the effective surcharge to the individual GCL shingle and, therefore, increase the FS of the cover system. It was found that one 0.45 m (1.5 feet) wide mid-slope trench can increase the FS by approximately 10 percent. Therefore, depending on construction and material costs, additional anchor trenches can be placed at strategic locations to further increase slope stability.

To verify whether a specific GCL can provide enough interface shear strength for slope stability, verification testing should be conducted using similar surcharge and design. For the proposed cover system, the interface shear strength requirements for a GCL and geosynthetic drainage blanket to meet an adequate FS for a 2:1 slope are shown in Figure 2. The lower and upper envelopes present a static FS of 1.5 and 1.8, respectively, for the overlying layers. It is estimated that at a static FS within this range, the anticipated displacement due to seismic loading would likely be small and, therefore, meet the dynamic FS requirement. The strength requirements could be achieved by any or a combination of the above improvement options.

To evaluate the slope stability under seismic conditions, a yield acceleration for each of the GCLs was calculated. The average residual shear strength from laboratory interface shear tests were compared to the static shear stress at the soil-GCL interface, assuming a 0.3 m (1-foot) thick final cover soil layer on a 2:1 slope with a vertical height of 12 m (40 feet) between benches.

Yield accelerations along critical potential slip surfaces for the various cover configurations ranged from 0.19g to about 0.69g. The Toyon Landfill is located in an area of high seismic activity, and peak ground accelerations at the site are likely to exceed 0.2g and may reach 0.6g. This indicates that FS against shallow sliding may be reduced to less than 1.0 under seismic loading conditions for most or all of the alternatives. A seismic analysis was conducted to estimate the nature and magnitude of shallow sliding due to earthquake loading conditions.

Pseudostatic analyses generally assume that a mass will fail if an earthquake-induced horizontal acceleration exceeds the yield acceleration from a limit equilibrium analysis. Since earthquake loading is transient in nature, a slope can be subjected to repeated pulses of acceleration and stress that may temporarily exceed the static failure strength along a failure surface. Each time the yield strength is exceeded, the mass defined by the failure surface will experience permanent deformation. Therefore, the total amount of deformation experienced by the mass depends on the number and frequency of transient pulses which produce stresses greater than the yield strength of the failure surface.

Seismic-induced permanent deformation along the interface of the final cover soil and the GCL has been evaluated using the approach summarized below:

1. A finite element mesh was generated for three representative cross-sections simulating maximum, intermediate, and minimum waste fill of the landfill, with elements representing the bedrock, refuse fill, and cover layer.

2. The finite element model FEADAM was utilized to estimate the distribution of stress within the landfill. The FEADAM output was used as input into QUAD-4 (discussed below).

3.   The finite element model QUAD-4 was utilized to evaluate the seismic response of the landfill to input ground motions. The dynamic response analysis uses strain-dependent properties for the fill to calculate time histories of acceleration within the landfill and at the cover soil-GCL interface. Two input ground motions, one representing a far-field event (San Andreas Fault) and one representing a mid-field event (San Gabriel Fault), were evaluated (I.T. Corporation, Toyon Landfill Final Closure and Postclosure Maintenance Plan, 1994).

4.   The seismic-induced deformation along the soil-GCL interface was calculated. The deformation analysis is based on comparison of calculated yield acceleration to induced acceleration determined using the dynamic response analysis.   Down-slope deformation was calculated at nodal points representing the GCL-cover soil interface using the measured short-term interface shear strength and the calculated response time histories from the QUAD-4 analysis.

Previous analyses indicated that the seismic response to the near-field event was less than that calculated for the far-field event and greater than the mid-field event. It is assumed that the probability of the near-field event occurring at the site is comparable to that for the far-field event. Because the far-field event would induce greater accelerations at the slope face, the near-field event was not evaluated.

The ground motions may be either attenuated or amplified, depending on location, depth of the refuse cell, predominant frequency, and other variables. Generally, peak acceleration response at the landfill surface was amplified from bedrock peak accelerations for the far-field event, with 0.27g peak bedrock accelerations producing peak accelerations as high as 0.5g at the landfill surface.

Peak surface accelerations in response to the mid-field event were generally equal to or less than the assumed bedrock acceleration of 0.44g. The latter result is consistent with recent measurements at the Operating Industries, Inc. (OII) landfill in Monterey Park, California, where peak accelerations of approximately 0.248g and 0.255g were recorded at the base and top of the landfill, respectively, during the Northridge earthquake (I.T. Corporation, Toyon Landfill Final Closure and Postclosure Maintenance Plan, 1994).

The QUAD-4 seismic response analysis provoked acceleration time histories at nodal points along the surface of the face slope. By double-integrating the acceleration histories at nodal points along the slope face, permanent displacements can be calculated.

Based on calculated yield accelerations for the various GCL products, the estimated range of seismic-induced shallow sliding corresponding to the assumed mid-field and far-field events are summarized below:

Configuration/Product	Calculated Displacement for Mid-Field Event (cm)	Calculated Displacement for Far-Field Event (cm)
Claymax 500 SP	0.0 to 3.1	0.0 to 15.2
Gundseal	0.0 to 9.1	6.1 to 54.9
Bentomat	0.0 to 6.1	3.1 to 36.6
DNW Bentomat	Negligible	Negligible

The analyses indicate that maximum displacements are likely to occur near the abutments on the upper portions of the slope. Displacements corresponding to a near-field event (i.e., a magnitude 6.75 event on the Richter Scale along the Santa Monica Fault) were not evaluated, but are expected to be between those calculated for the mid- and far-field events. While a probabilistic seismic analysis has not been performed, it is reasonable to assume that during the design life of the site, several episodes of ground motion similar to that assumed for the mid-field event may occur. The probability of the site being subjected to more than one episode of ground motions as strong as those modeled by the assumed far-field event is thought to be low.

Deformation will probably occur as tension cracks at or near the top of individual benched slopes, which will require cosmetic repair. The uppermost soil-GCL interface represents the weakest layer, and, therefore, the displacements should occur between the cover soil and GCL. The integrity of the GCL should not be compromised by the potential displacement of cover soil. Seismic-induced slippage of the cover soil may also cause localized damage to components of the gas collection, irrigation, and surface drainage systems. Seismic-induced deformations within the range calculated for any of the products are considered to be manageable and will be planned and budgeted for as a maintenance issue during the postclosure period.

Long Term Creep Tests
The final cover system at the front-face slope will experience a constant shear stress due to its inclination. Based on the proposed 2:1 final grade, this constant shear stress is expected to be about 45 percent of the overburden weight. To evaluate the potential impact of this shear stress on the GCL layer, long-term creep tests were performed.

The GCL tested is the double nonwoven (DNW) GCL that had the highest shear strength of all GCLs studied to date.

Sample preparation and the testing apparatus for the long-term tests were identical to that for the short-term tests. A normal stress of 624 kgf/m^2 (128 psf) was applied to each specimen. After the test specimens were soaked for 24 hours, an initial displacement reading was taken from a dial gauge attached to the back of the shear box. A dead weight was then placed on a cable assembly attached to the shear box to apply a constant shear stress to the test specimen. Displacement readings were recorded at elapsed time intervals of 1, 2, 4, 8, 15, 30, 60, and 120 minutes and thereafter every 24 to 72 hours.

Five different constant shear stresses ranging from about 30 to 80 percent of the average short-term residual interface shear strength were initially selected (one shear stress level applied on each of the five test devices). The shear stress on one test specimen was later increased to a level corresponding to about 104 percent of the average residual strength. Data for the long-term tests were collected for at least 1,000 hours of elapsed time or until the rate of displacement achieves a steady-state value, i.e., the slope of the displacement versus time plot is constant. Figure 3 presents results from two such tests.

It is noted that for the 2:1 slope and the proposed 0.3 m (1 foot) thick final cover configuration proposed for the Toyon landfill, the static shear stress at the soil-GCL interface will be approximately $1.95 kgf/m^2$ (0.4 psi). Based on the results of the creep tests on the DNW Bentomat, down-slope creep on the order of $1.27x10^{-5}$ to $2.54x10^{-5}$ cm/hour ($5 x 10^{-6}$ to $10 x 10^6$ inch/hour), or about 0.13 to 0.25 cm/year (0.05 to 0.1 inch/year) should be anticipated. This amount of creep is expected to be manageable through routine postclosure maintenance. No significant adverse impact is expected from this magnitude of deformation.

Integrated Final Cover
The stability of the final cover, however important, is only one of the vital components of final closure design. The closure design must also address several important issues, two of which are leachate control and landfill gas collection and migration control.

Leachate Control
The Toyon Canyon Landfill is located in an area of Griffith Park, Los Angeles, predominated by fractured sandstone bedrock and artesian springs. These springs, combined with the lack of an impermeable final cover, have historically produced numerous small leachate seeps on the front face.

As the site was constructed in 1957, there are not any subsurface leachate control structures. However, after leachate seeps were observed in the past and to prevent the degradation of the groundwater aquifer, a series of horizontal collector pipes were installed as french drains which collected leachate along the benches before routing it to the sanitary sewer system. Also, a barrier cut-off wall was installed at the toe of the landfill which intercepted leachate before it mixed with groundwater, routing it to the sewer system. The ubiquitous nature of the leachate seeps on the front face dictated a more comprehensive approach in leachate management. This was accomplished during final closure design as discussed below.

A drainage layer, though not specifically called for, was incorporated into the final cover design, and exists between the foundation layer and the hydraulic barrier layer.

The drainage layer consists of a high density polyethylene (HDPE) geonet, surrounded by two layers of a non woven geofabric. This drainage layer is designed and placed to collect any seepage from the refuse fill that may penetrate through the foundation layer (Figure 4) before coming into contact with the hydraulic barrier layer. This liquid or leachate (called so due to its contact with refuse) is then routed down to the bottom of the slope (toe of slope) through the geonet where it enters a collection trench drainage system (french drains with 20 cm, 8 inch, PVC collector pipes) located along every bench. The leachate is then routed further down-slope to the toe of the front face via 20 cm (8 inch) PVC collection pipes, and ultimately to the sanitary sewer for disposal.

The 20 cm (8 inch) piping system is oversized for the site (anticipated reduction in overall leachate generation after placement of the final cover to a maximum rate of 57 liters, 15 gallons, per minute), but allows for ease of maintenance. Maintenance of the existing leachate system indicates a carbonaceous build-up over time with must be forcibly removed on a quarterly basis. Installation of 20 cm (8 inch) pipes allows for maintenance equipment to be easily used when cleaning the pipes.

This design will eliminate any front face leachate seeps, by internally managing liquid flow through the underground leachate collection and removal system (LCRS).

Landfill Gas Collection and Migration Control
At Toyon Canyon Landfill, a privately-owned LFG collection system is operated for electrical generation. As such, the owners/operators of the energy conversion facility desire optimum plant efficiency, which translates into collecting only the rich portion of LFG, and minimizing or eliminating the leaner LFG at or near the refuse boundary.

Pacific Energy collects the gas on-site, where it is combusted to generate 8.9 megawatts of electricity. The collection rate is 68 to 74 m^3/min (2400-2600 scfm) via 200 landfill gas wells. The gas collection system is limited to the top deck and benches 6 through 10. This system is designed to optimize gas collection efficiency for revenue generation rather than migration control. Consequently, surface emissions exceeding the compliance limits occur monthly on benches 1 through 5 and at the refuse perimeter and interfaces.

Additionally, off-site migration typically exceeds the 5% methane compliance limit in 20 of the 34 perimeter landfill gas migration control probes. Regulations enforced by the CIWMB are set forth in 14 CCR, which stipulate a five percent methane concentration limit at the landfill property boundary. Landfill gas migration control has been a problem at the Toyon site, due in part to the operation of the LFG collection system and the placement of LFG migration probes at or near (3-6 m, 10-20 feet) the refuse boundary. One-half of the LFG migration solution is to more sensibly relocate the LFG probes, but the other half of the solution must be developed at that refuse boundary or perimeter.

The final cover system is expected to increase overall LFG efficiencies and aid in controlling off-site migration.   Surface emissions are anticipated to be eliminated throughout the landfill by placement of the low permeability hydraulic barrier. Also, the low permeability cover will allow increasing gas vacuum to be applied at the landfill's perimeter, aiding in controlling off-site migration.

On the slopes, the landfill gas system will also be connected to the leachate control system at the cleanouts.   A low flow/vacuum will be applied across the entire front face through the lateral leachate pipes and into the geonet system, thereby eliminating any surface emissions.

CCR Title 14 requires landfill gas monitoring for 30 years after closure of the landfill. For the Toyon Landfill, gas monitoring would be required until the year 2015. Landfill gas will continue to be controlled at the energy recovery plant during this time period, with LFG combusted in the five internal combustion engines. The Bureau recognizes that gas collection will begin decreasing and that energy generation will eventually require supplemental fuel augmentation, thereby limiting the economic benefits of the energy generation facility. At this point, when operating expenditures exceed revenues, Pacific Energy will begin removing one or more of the internal combustion engines and eventually shut-down the facility. The Bureau is expected to receive a two-year advance warning of the energy facility shut-down, and will either acquire the energy facility for LFG environmental control or design and install a flare system.

Conclusion
Final closure design is perhaps the greatest challenge to landfill engineers.   Its complexity transcends operating landfill design projects, especially true within the realm of Subtitle D regulations and any state and local requirements.   California regulations are among the most stringent in the nation.   The landfill engineer must recognize and understand the interrelationships that exist between the final cover - its static and seismic stability, leachate control, and LFG collection and migration control in order to overcome their design challenges. This understanding is important for a successful and economic design, one that once constructed, will stand the test of time (including seismic activity).

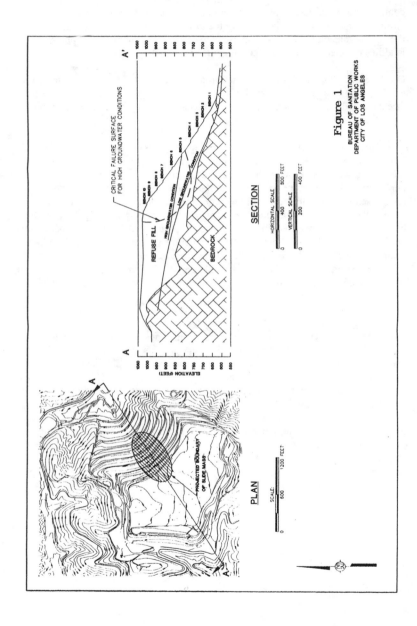

SECTION

PLAN

CRITICAL FAILURE SURFACE
FOR HIGH GROUNDWATER CONDITIONS

REFUSE FILL

BEDROCK

Figure 1

BUREAU OF SANITATION
DEPARTMENT OF PUBLIC WORKS
CITY OF LOS ANGELES

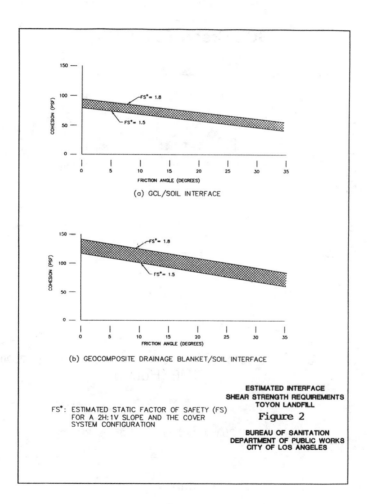

FS*: ESTIMATED STATIC FACTOR OF SAFETY (FS)
FOR A 2H:1V SLOPE AND THE COVER
SYSTEM CONFIGURATION

ESTIMATED INTERFACE
SHEAR STRENGTH REQUIREMENTS
TOYON LANDFILL

Figure 2

BUREAU OF SANITATION
DEPARTMENT OF PUBLIC WORKS
CITY OF LOS ANGELES

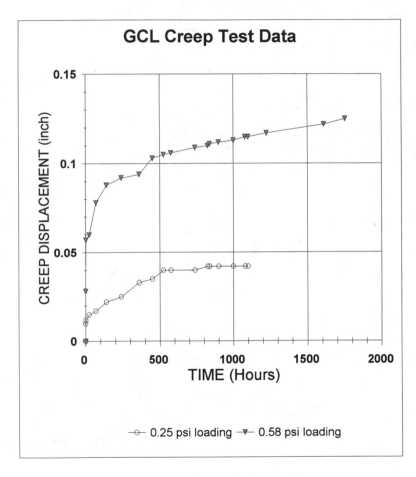

Figure 3

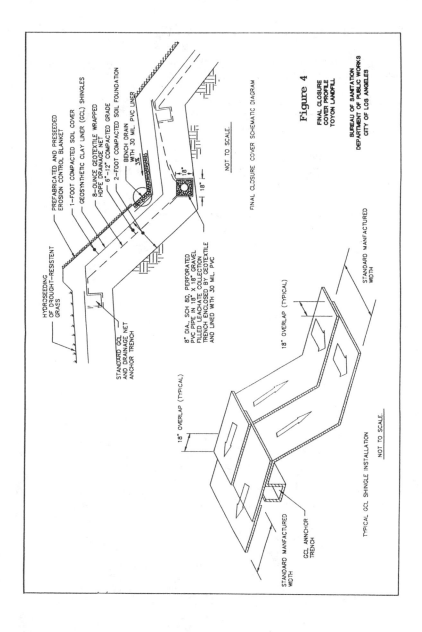

HYDROSEEDING OF DROUGHT-RESISTENT GRASS

PREFABRICATED AND PRESEEDED EROSION CONTROL BLANKET

1-FOOT COMPACTED SOIL COVER

GEOSYNTHETIC CLAY LINER (GCL) SHINGLES

8-OUNCE GEOTEXTILE WRAPPED HDPE DRAINAGE NET

6"-12" COMPACTED GRADE

2-FOOT COMPACTED SOIL FOUNDATION

BENCH DRAIN WITH 30 MIL. PVC LINER
3%

STANDARD GCL AND DRAINAGE NET ANCHOR TRENCH

8" DIA., SCH 80, PERFORATED PVC PIPE IN 18" X 18" GRAVEL FILLED LEACHATE COLLECTION TRENCH ENCLOSED BY GEOTEXTILE AND LINED WITH 30 MIL. PVC

18"

18"

NOT TO SCALE.

FINAL CLOSURE COVER SCHEMATIC DIAGRAM

18" OVERLAP (TYPICAL)

18" OVERLAP (TYPICAL)

STANDARD MANFACTURED WIDTH

STANDARD MANFACTURED WIDTH

GCL ANNCHOR TRENCH

TYPICAL GCL SHINGLE INSTALLATION

NOT TO SCALE.

Figure 4

FINAL CLOSURE
COVER PROFILE
TOYON LANDFILL

BUREAU OF SANITATION
DEPARTMENT OF PUBLIC WORKS
CITY OF LOS ANGELES

REFERENCES

Barclays California Code of Regulations, Titles 14 and 23.

Dobrowolski, Jeffrey G. and Pano, Constantin, "Horizontal Collectors for Landfill Gas Collection and Migration Control", March, 1995.

I.T. Corporation, 1988, "Solid Waste Assessment Test (SWAT), Toyon Landfill," prepared for the Bureau of Sanitation, Department of Public Works, City of Los Angeles, June.

I.T. Corporation, 1994, "Final Closure and Postclosure Maintenance Plan for Toyon Canyon Sanitary Landfill," prepared for the Bureau of Sanitation, Department of Public Works, City of Los Angeles, September.

LANDFILL FINAL COVERS: A REVIEW OF CALIFORNIA PRACTICE

Krzysztof S. Jesionek[1], M.ASCE and R. Jeffrey Dunn[2], M.ASCE

Abstract

This paper presents a summary of results of a survey of landfill cover systems being used at California landfills. The survey was completed as an early phase in a two-year study of performance criteria for landfill covers completed by GeoSyntec Consultants for the California Integrated Waste Management Board. Conclusions from the survey and recommendations related to modifying the existing cover regulations are also presented.

Introduction

GeoSyntec Consultants was commissioned in 1991 by the California Integrated Waste Management Board (CIWMB) to develop performance criteria for landfill covers for municipal solid waste landfills (MSWLs) in California. As part of this two-year study, a survey was distributed to 352 owners/operators of active landfills in California, out of which 122, or 35%, responded. The purpose of the survey was to gather information on various systems and materials being used as daily cover, intermediate cover, and final cover in the State of California and on modifications to cover practices felt to be appropriate by the landfill operators. This paper deals only with final covers. The survey predated implementation of federal landfill requirements in Part 258, Title 40 of the Federal Code of Regulations (Subtitle D), although conformance of California regulations to Subtitle D closure requirements was also part of the overall project.

[1]Manager, Solid Waste Group, GeoSyntec Consultants, 1600 Riviera Avenue, Suite 420, Walnut Creek, California, 94596

[2]Principal, GeoSyntec Consultants, 1600 Riviera Avenue, Suite 420, Walnut Creek, 94596

The questionnaire was divided into five parts and requested the following:
* general site information;
* information on landfill operations including waste compaction equipment, waste lift height, number of passes of compaction equipment;
* information on interim cover including daily and intermediate covers;
* information on final cover including type and construction methods of the three basic layers that are required for a final cover system in California (i.e., foundation layer, barrier/infiltration layer, and vegetation/erosion layer), any other layers installed as part of the final cover system, and final cover construction quality assurance (CQA); and
* comments on existing requirements related to landfill covers and recommendations for modifications.

Landfill Covers

Landfill covers at MSWLs serve a variety of functions depending on the type of cover. Generally, two types of landfill covers are defined in California: interim cover (daily and intermediate) and final cover. Interim cover is spread on the entire surface of the active face of the sanitary landfill in order to control vectors, fire, water infiltration, erosion, and to prevent unsightliness. Interim cover is not the subject of this paper. The main objectives in designing a final cover system include the following [EPA, 1991]:
* minimize infiltration of water into the waste, if the waste is to be kept dry;
* promote surface drainage and maximize runoff;
* control gas migration; and
* provide a physical separation between waste and humans, plants, and animals.

In current practice, the primary function of a final cover is to minimize infiltration into the waste and consequently reduce the amount of leachate being generated. The principal factor affecting leachate generation is infiltration of precipitation through the final cover and into the underlying waste. The rate of infiltration varies and depends on the material used in the cover (e.g., earth or geosynthetics), the cover's relevant physical properties (e.g., hydraulic conductivity, moisture retention capability), the long-term integrity of the cover, and the climate in which the landfill is located [Jesionek et al., 1995]. For example, an earth material's properties may change with time due to erosion, differential settlement, freeze-thaw cycles, and desiccation. EPA [1991] suggested that in the case that the waste is continuing to settle, e.g., as a result of decomposition, it may be prudent to place a temporary cover on the waste and wait for settlement to take place prior to constructing the final cover system. The appropriate waiting period would vary from site to site, but generally would be measured in years. Unfortunately, owners and operators are generally not allowed by current regulations to wait to complete final landfill closure.

The potential components of a final cover system for a MSWL are shown on Figure 1 [Jesionek et al., 1995]. As recognized by the state and federal regulations, all five components of the final cover system shown are not required. A final cover system may include just two (as required by the federal regulations) or three (as required by the California regulations) of the five possible components shown on Figure 1. The United States Environmental Protection Agency (EPA) has established the minimum requirements for MSWLs [EPA, 1992]. When the prescriptive liner system is used in a waste disposal unit, EPA requires that the final cover, as a minimum, must include the components shown on Figure 2. Figure 3 shows a typical profile of the final cover system required by California. Jesionek et al. [1995] provide a comprehensive evaluation of different types of materials used in the MSWL final cover systems.

Regulatory Requirements

The federal and California regulatory requirements for landfill final covers are included in Subtitle D, Title 14 of the California Code of Regulations (CCR) (Title 14), and in Chapter 15, Title 23 of the CCR (Chapter 15).

Prescriptive requirements of Subtitle D require that the final cover system be comprised of an erosion layer underlain by an infiltration layer. The erosion layer should consist of a minimum 150 mm of earthen material that is capable of sustaining native plant growth. The infiltration (barrier) layer is to be constructed of a minimum 460 mm of earthen material that has a hydraulic conductivity less than or equal to the hydraulic conductivity of a bottom liner system or natural subsoils present, or a hydraulic conductivity no greater than 1×10^{-5} cm/s, whichever is less.

The prescriptive requirements of Chapter 15 require that a final cover system include at least a 600-mm thick foundation layer, a 300-mm thick barrier layer, and a 300-mm thick vegetation (erosion) layer (Figure 3). The foundation layer should be composed of soil, contaminated soil, incinerator ash, or other waste materials, provided that such materials have appropriate engineering properties. The barrier layer should consist of soil containing no waste or leachate and compacted to attain a hydraulic conductivity of either 1×10^{-6} cm/s or less, or equal to the hydraulic conductivity of any bottom liner system or underlying natural geologic materials, whichever is less. The vegetation (erosion) layer should be comprised of soil containing no waste or leachate.

Survey Results

A brief description of some of the major points found from the survey results is presented herein. Additional details are presented by GeoSyntec Consultants [1994].

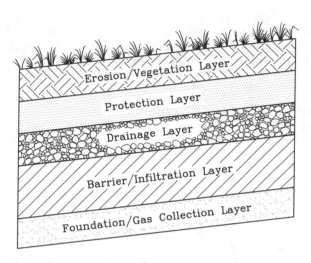

Figure 1. Final cover system components

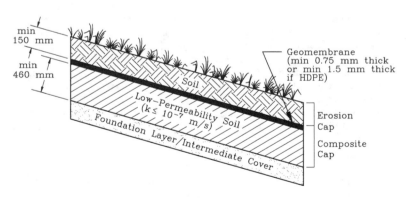

Figure 2. EPA recommended minimum final cover system for MSW landfills

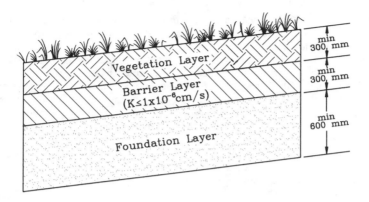

Figure 3. Typical California Chapter 15 final cover system

General. The returned survey questionnaires indicated that permitted areas of active landfills in California ranged from 0.4 to 612 hectares (ha) with the average permitted area of 61 ha. Projected dates of landfill final closure varied from 1994 to 2130.

Thirty-four respondents to the survey (28%) provided sketches of cross-sections for their landfills' final covers. Of those who provided the cross-sections, 50% indicated that their final cover configuration follows the prescriptive standard of Chapter 15, i.e., 300 mm of vegetative layer material, 300 mm of low permeability material, and 600 mm of foundation layer material. At three landfills, geosynthetic membranes formed a composite barrier layer with the low permeability clay layer. The remaining configurations were variations of the prescriptive standard. These included 460 mm or 600 mm of the vegetative or low permeability layer instead of 300 mm required for each layer.

The respondents listed different types of waste compaction equipment used at landfills in California. However, the majority of landfill operators use compactors and/or dozers with operating weights ranging from 7,260 to 45,360 kg. The reported number of passes ranged from 2 to 10 with the typical and average numbers of 3 and 4, respectively. Reported lift heights ranged from 300 mm to 9 m before compaction and between 100 mm and 7 m after the waste has been compacted. Some respondents indicated that additional compactive effort is being placed on waste prior to placing the final cover.

Final Covers. The majority of responding landfill owners/operators used soil (including contaminated soil) as the foundation layer material. Soil was also indicated as the preferable material for use for foundation layer in the future (77%). Although allowed by the regulations, only one respondent indicated the use of waste material (incinerator ash) for the foundation layer. Reported thicknesses of foundation layers varied from 150 to 900 mm. The survey also showed that almost a third of the landfills did not install a foundation layer during construction of the final cover system. Reported soils used for foundation layer varied from silty sand to clayey loam, with reported hydraulic conductivities ranging from 1×10^{-3} to 1×10^{-7} cm/s. It should be noted that foundation layer hydraulic conductivity is not a commonly tested parameter, but it can be important to performance evaluation of a cover. Compaction of the foundation layer material was reported to be accomplished mainly with sheepsfoot type compactors. Operating weights of these compactors ranged from 13,600 to 45,300 kg. The average and typical number of passes by a compactor to compact the foundation layer material varied between 4 and 6, respectively. Typically, the foundation layer material was compacted to 90% of the maximum dry density, as evaluated by ASTM Standard D 1557 (modified Proctor test method), with typical moisture contents ranging from 0 to 3% above optimum moisture content (ASTM D 1557). The typical foundation layer material lift height was about 200 mm before compaction. The survey respondents did not indicate the use of geogrids or other types of soil-reinforcement materials in foundation layers.

According to the survey, most of the respondents have used natural clay soils to construct final cover barrier (infiltration) layers. However, only one half indicated clay as being the preferable material to be used for the barrier layer. One respondent indicated the application of a geosynthetic clay liner (GCL), and three others the usage of high density polyethylene (HDPE) and very low density polyethylene (VLDPE) for barrier layers. About 60% of the respondents said that clays used at their sites have been imported. Reported thicknesses of clay barriers ranged from 300 to 600 mm with a typical thickness of 300 mm. Reported hydraulic conductivities of the clay barrier varied from 1×10^{-5} to 1×10^{-8} cm/s. However, the predominant value of the hydraulic conductivity was reported to be 1×10^{-6} cm/s. Compaction of clay, used for the barrier layer construction, has been accomplished mainly with sheepsfoot type compactors. A typical barrier layer has been compacted to 90% of the maximum dry density, as determined by ASTM D 1557, with typical moisture contents ranging from 0 to 3% above optimum moisture content (ASTM D 1557). Required compaction was typically achieved with 6 passes of the compaction equipment. A 200 mm typical lift height, before compaction, was reported.

Vegetative (erosion) layers, as depicted by the survey respondents, consist almost entirely of soils native to the respective sites. The average thickness of the layer was 430 mm, with a minimum of 150 mm and maximum of 1.2 m. According to the survey, a hydraulic conductivity of the vegetative layer in the California landfills ranged from 1×10^{-3} to 1×10^{-8} cm/s. Although the testing method for the layer's hydraulic conductivity was not requested in the survey questionnaire, the data provided indicated the lower end range of hydraulic conductivity of 1×10^{-8} cm/s of the erosion (vegetation) layer material. It is noted that this low hydraulic conductivity may not be compatible with vegetation growth.   Although the majority of the respondents indicated that they did not use or desire to use materials other than soils for construction of the vegetative (erosion) layer, sludge, sludge amended soils, and compost were among those materials which the landfill owners/operators would consider to use.

Other Layers.   The section of the questionnaire inquiring about other layers in the final cover configuration, such as drainage layer, filter layer, biotic layer, or gas vent layer, elicited only four responses (3%).   Three of these responses, however, came from the same landfill owner/operator responding for three different landfills.   The operator apparently used weathered granite to construct drainage and gas vent layers.

Comments.   The final part of the questionnaire provided an opportunity for landfill owners/operators responding to the survey to comment on any subject which might be helpful to this study.   A third of the respondents commented on different aspects of landfill operations in California, not always necessarily covered by the survey.   These comments included the need for adopting a final cover configuration close to the time of final closure, rather than earlier in the life of an operating landfill.   These comments may be related to a general feeling expressed by landfill owners/operators that regulatory requirements are very likely to change in the future as a result of studies on landfill cover materials and the introduction of new materials such as geosynthetic clay liners.   In a related matter, one landfill owner/operator complained about "outdated and too clumsy" procedures for evaluating landfill covers.

Conclusions Developed From Survey

Final cover systems at MSWLs serve a variety of functions for both new waste disposal units as well as older units being retrofitted with a modern cover. Because of the variety of waste characteristics, available materials, and site conditions, cover systems must be tailored to the specific requirements of each particular project.   Although prescriptive final covers stipulated by regulations may offer useful guidelines, it is important for landfill owners/operators, designers, and regulatory officials to understand that the components of a final cover system, and the best materials to use for each component, are not the same for all sites.   Each site should be evaluated on a case-by-case basis.

The typical final cover configuration at most MSWLs in California follows the Chapter 15 prescriptive standard.   This configuration (also known as 2-1-1 configuration), consisting of a minimum 600-mm thick foundation layer overlain by a minimum 300-mm thick barrier layer and a minimum 300-mm thick vegetation layer.   The survey results indicate, that although clay remains a predominant material for construction of the barrier (infiltration) layer in the final cover, a number of respondents indicated geosynthetic clay liners (GCL) and geomembranes as preferred materials. This preference is consistent with practice in landfill closure in many parts of the United States where geosynthetics are replacing clay materials owing to potential for clay degradation under drying or freezing conditions or limited availability of clay materials

Figure 3 shows a typical landfill final cover used in California. The profile has the advantages of simplicity in concept and construction, as well as low cost. The main problems with this design are in maintaining vegetative growth during drought conditions, lack of protection of the barrier layer from wet/dry cycles that can crack the soil, the relatively higher allowable hydraulic conductivity (often 1×10^{-6} cm/s) of the barrier layer, as compared to current liner requirements, lack of a gas collection layer, and lack of redundancy in the profile. The Chapter 15 final cover system will function with some effectiveness, but will not provide as high a level of infiltration reduction and landfill gas control as more complex cover sections.

The main advantages of utilizing the Chapter 15 cover configuration include:
* construction materials are often available locally;
* topsoil stores water that has infiltrated; plants return much of this water to the atmosphere via evapotranspiration;
* plant roots reduce soil erosion and stabilize the topsoil;
* barrier layer enhances soil moisture storage in topsoil and helps to promote evapotranspiration of water;
* cover is relatively easy to construct; and
* cover is relatively inexpensive.

Major disadvantages of the Chapter 15 cover are as follows:
* requires adequate precipitation to support growth of vegetation on topsoil;
* establishing growth of vegetation may be difficult if weather conditions are unfavorable immediately after seeding;
* excessive wind or water erosion may occur prior to establishment of vegetative cover or if vegetative cover becomes thin;
* little protection of barrier layer from desiccation-induced cracking due to thinness of topsoil;
* thinness of barrier layer makes it vulnerable to damage or construction defects;
* no redundancy in surface or barrier layer; and
* no gas collection layer.

Recommendations

Due to the variety of waste characteristics, available materials, and site conditions, final cover systems must be tailored to the specific requirements of each particular site. Prescriptive final cover configurations should be viewed as guidelines only. In most instances, simply following prescriptive standards will not provide the best solution for minimizing leachate generation. It is imperative for landfill owners/operators, designers, and regulatory officials to understand that the components of a final cover system, and the best materials to use for each component, should not be generalized for all sites. Thus, a landfill final cover system, including a variety of available materials, should be designed to meet the performance objectives of landfill covers and not simply mimic the prescriptive design. Many types of non-prescriptive covers can and do meet performance objectives. It is the authors' and others [e.g., Daniel and Richardson, 1995] opinion that design engineers and regulatory agencies, in order to reduce potential for ground-water contamination, consider geomembranes or GCLs, or both, for all MSWL covers, regardless of whether a composite liner system has been installed in the landfill liner or not.

To aid in the final cover system selection, GeoSyntec Consultants [1994] developed a flow chart which compiles information generated for typical cover configurations, setting, design elements, and design methods. The flow chart should aid regulatory agencies reviewing proposed cover designs and landfill owners/operators preparing the design. Using the flow chart, one should be able to quickly see what issues should be considered in the design. A portion of the flow chart, for design of the barrier layer of the final cover system, is presented on Figure 4 [Jesionek et al., 1995].

The flow chart has been developed for a typical final cover system design. Therefore, it provides general information and guidelines for such a system. The landfill final cover design process should begin with a review of local, state and federal regulations. This review will indicate minimum standards and permissible configurations for the cover. Site characteristics, such as climate, available soil cover materials, waste properties, seismicity, etc., should be collected and analyzed. This analysis will help in identifying potential cover materials that can, for example, exhibit favorable properties for foundation and barrier layer, and supporting vegetation. The anticipated post-closure activities at the site should be considered when evaluating and selecting cover components. Post-closure requirements may involve development of commercial facilities, park or simply aesthetic landscaping which could affect the thickness and characteristics of the components of the final cover. Experience may play an important role in evaluating cover systems that have performed satisfactorily under similar climatological and other conditions. Potential cover materials, which are available through manufacturers and aggregate suppliers, play a major factor in expanding the selection process to fulfill a design or performance requirement.

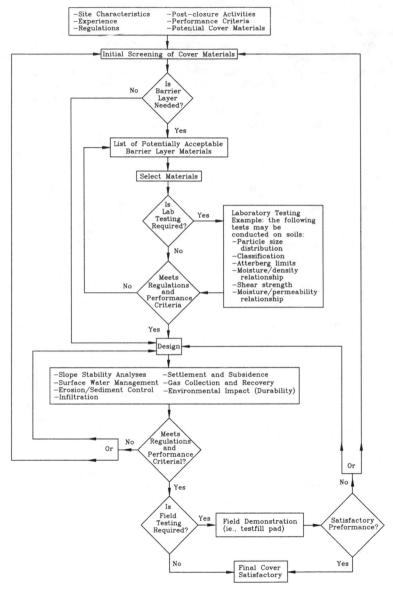

Figure 4. Flow chart for selection of final cover system components [after Jesionek et al., 1995]

Finally, the performance criteria should be used as a preliminary tool in the selection or review process of possible cover components identified above. This will help in identifying one or more cover cross-sections that can be suitable as final covers. Individual components of selected final covers would then be evaluated to see if appropriate laboratory testing is required. Each component and the final cover configuration would then be designed to meet regulations and performance criteria. Examples of some typical laboratory tests, design methods, and design elements, if required for the particular layer of a final cap system or for the entire configuration, are also listed on the flow chart. Ultimately, a field demonstration of the final cover selected may be required to validate its performance.

Acknowledgment

The authors wish to acknowledge the support of the California Integrated Waste Management Board which provided the research support for the work described herein.

References

Daniel, D.E., and G.N. Richardson, 1995. "The Role of Geomembranes and Geosynthetic Clay Liners in Landfill Covers", *Geotechnical Fabrics Report*, Vol. 3, No. 1, January/February 1995, pp. 44-49.

Environmental Protection Agency, 1991. "*Design and Construction of RCRA/CERCLA Final Covers*", EPA/625/4-91/025, Washington, D.C., 145 p.

Environmental Protection Agency, 1992. "*Final Rule Corrections: Solid Waste Facility Disposal Criteria*", Federal Register, Vol. 57, No. 124, pp. 28626-28629.

GeoSyntec Consultants, 1994. "*Landfill Cover Material Identification and Evaluation*", A Report Prepared for the California Integrated Waste Management Board, 222 p.

Jesionek, K.S., R.J. Dunn and D.E. Daniel, "*Evaluation of Landfill Final Covers*", Proceedings of the 5th International Landfill Symposium, Sardinia, 2-6 October 1995, 25 p. (Accepted for Publication).

GEOTECHNICAL MONITORING OF THE OII LANDFILL

by Peter K. Mundy[1], M. ASCE, John P. Nyznyk[2], Seyed Ali Bastani[1, 4],
William D. Brick[1], Jocelyn Clark[1], and Roy Herzig[3]

ABSTRACT

To help assess slope stability and settlement for the design of a final cover system, in December 1987 the U. S. Environmental Protection Agency (EPA) established a quarterly geotechnical monitoring program at the Operating Industries, Inc. (OII) Landfill Superfund site. Geotechnical instrumentation included inclinometers, extensometers, piezometers, and surface monuments. The results of seven years of quarterly monitoring are presented and discussed in this paper. The general trend of movement for the OII Landfill slopes is down and out-of-slope. These movements have occurred at a fairly constant rate. Overall, the displacement vectors of movement show that the landfill is, in effect, shrinking. This field performance data has significant potential value regarding evaluation of expected long-term landfill settlements, slumping and lateral displacements. Moreover, field observations may also provide valuable insights regarding the mechanisms responsible for observed settlements and lateral displacements, which could be useful in making projections of likely future displacements.

INTRODUCTION

The Operating Industries, Inc. (OII) Landfill is a Superfund site located approximately 10 miles east of Los Angeles, within the City of Monterey Park and adjacent to the City of Montebello, California. Throughout its operating life (1948 to 1984), residential, commercial, industrial, liquid, and various hazardous wastes were disposed at the landfill. In 1987, the U. S. Environmental Protection Agency (EPA) established a geotechnical monitoring program at the site in response to indications of slope and cover movement. The ongoing geotechnical monitoring program is under the direction of EPA. The program objectives were to determine the magnitude and

---

1. CDM Federal Programs Corporation, 100 Pringle Avenue, Suite 500, Walnut Creek, California 94596
2. Camp Dresser & McKee, 100 Pringle Avenue, Suite 300, Walnut Creek, California 94596
3. US Environmental Protection Agency, 75 Hawthorne Street, San Francisco, California 94105
4. Bing Yen & Associates, Inc., 17701 Mitchell North, Irvine, California 92714

direction of slope movements so that an assessment could be made of the potential for slope failure under both gravity (static) and seismic loading. This paper presents a compilation and interpretation of field performance data collected from the OII Landfill over the last seven years.

BACKGROUND

The OII Landfill encompasses approximately 190 acres. The Pomona Freeway divides the site into a 45-acre North Parcel and 145-acre South Parcel as shown in Figure 1. The OII Landfill is founded on Pliocene- to Pleistocene-aged sediments of the Pico Unit and the Lakewood and San Pedro Formations (EPA, 1994). The Pico Unit consists of alternating sandstones, sandy shales, clayey shales, and siltstones. The San Pedro Formation consists of coarse-grained sandstones and conglomerates with interbedded siltstone and clay lenses. The Lakewood Formation is a fluvial/ alluvial fan association unit that overlies the San Pedro Formation and, locally, the Pico Unit. Regionally, fine-grained sediments comprise between 40 and 80 percent of the total deposits in the Lakewood Formation. Coarse sandstones and conglomerates are typical of the basal deposits.

Landfilling operations began in 1948 with the filling of an existing natural drainage canyon — an area now occupied by a portion of the Pomona Freeway and north-central portions of the South Parcel. There is no evidence indicating subgrade preparation or installation of a liner system before the placement of waste. The OII Landfill was constructed by cut-and-cover filling operations. Whenever sand and gravel rich deposits were encountered, quarrying operations supplemented the cutting operations. By 1958, South Parcel filling had occurred over an approximate 25-acre portion of the central part of the parcel to an elevation of approximately 500 feet (mean sea level).

This early filling established a steep north-facing fill slope; subsequent filling expanded the landfill to the south, east, and west by sequentially excavating and covering wastes with native soils. Some of the excavations extended below the toe of the original landfilled areas and had steep side slopes. As the excavations were filled, relatively young waste began to be placed on top of sloped intermediate covers of the original landfilled areas. The process of expanding the landfill radially beyond the originally landfilled area created complex bottom, side slope, and interior side slope geometry, and is herein referred to as the build-out geometry of the landfill.

Slopes of the OII Landfill range up to 230 feet high and have narrow benches at several levels. The average steepness over the total slope height ranges from approximately 2:1 to 3:1 (horizontal to vertical). Intermediate slopes are up to 100 feet high and up to 1.5:1 in steepness.

Landfilling operations ceased in October 1984. In January 1984, before disposal operations ended, the State of California placed the OII Landfill on the California Hazardous Waste Priority List. That same year, EPA proposed that the OII Landfill be added to the federal National Priorities List of Superfund sites. In May 1986, the OII Landfill was placed on the National Priorities List.

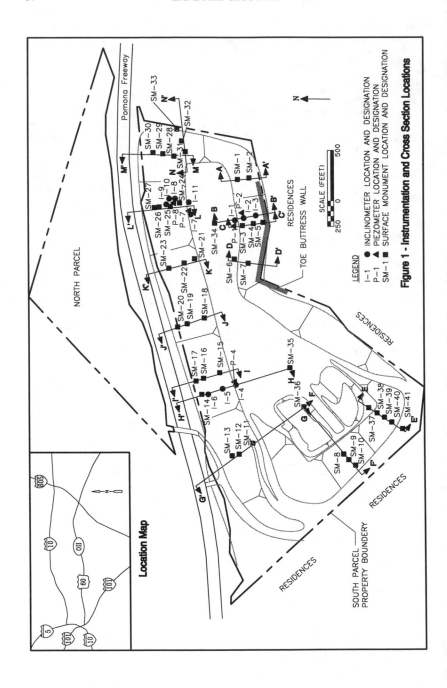

Figure 1 - Instrumentation and Cross Section Locations

LEGEND

I-1 ● INCLINOMETER LOCATION AND DESIGNATION
P-1 ▲ PIEZOMETER LOCATION AND DESIGNATION
SM-1 ■ SURFACE MONUMENT LOCATION AND DESIGNATION

SCALE (FEET)
250 0 500

Location Map

The OII Landfill is presently covered with soils that vary in thickness from 1 to 30 feet and consist primarily of silty clay to silty sand. Surface cracking, depressions, and erosional features are evident at various locations around the landfill. Features that appear as bulging and slumping are also evident. Refuse is sometimes exposed at these and other locations. As part of ongoing site maintenance, some of these features have been altered by regrading.

In 1987, EPA constructed a toe buttress of mechanically stabilized earth along the southeastern boundary of the landfill above the residences. The toe buttress is approximately 15 feet high and 1,500 feet long, and was constructed as an emergency response to observed slope movements in this area.

Also, in December 1987 under the direction of EPA, a geotechnical monitoring program was initiated in response to indications of slope movement. The primary objective of the program was to collect data necessary for determining the nature and extent of slope and cover movement. To meet this objective, EPA installed a network of geotechnical instrumentation in 1987. This installation included 36 surface settlement monuments, 11 inclinometers (incorporated into the inclinometers were two types of extensometers, Sondex and telescoping), and 5 piezometers. To monitor clay stockpiling operations, five additional surface monuments were installed in 1991. The instrumentation is arranged to form 14 cross-sections as shown in Figure 1. The selection and rationale for the location of each instrument (as well as a detailed discussion of the data compilation and interpretation) are presented in EPA 1995a.

GEOTECHNICAL MONITORING RESULTS

Surface Movements

A National Geodetic Survey third order, Class I survey was performed each monitoring session to obtain location and elevation of each surface monument, inclinometer, and piezometer. A network of survey control points were established off the Landfill and were based on the National Geodetic Vertical Datum (Mean Sea Level) and the California Coordinate System.

To assess the survey data quality, a detailed deformation analysis was performed (EPA, 1995b). The analyses, which utilize appropriate deformation analysis methodology, involved the re-adjustment and analysis of all possible survey events, and are characterized by the rigorous propagation of all statistical information using least squares adjustment techniques. The Iterative Weighted Similarity Transformation (IWST) methodology was utilized to verify the stability of the reference points. Based on the outcome of the IWST's, combined multi-epoch least squares adjustments were performed for the estimation of the final displacement vectors, and the associated 95 percent confidence regions. This methodology yields unambiguous results with rigorous statistical qualifiers for all "displacements." The results of these analyses were used in the plotting and interpretation of the geotechnical monitoring data.

Horizontal Surface Movements: The horizontal component of surface movement ranged from about 0.25 to 3.9 feet over the period from December 1987 through October 1994 (Figure 2). These resultant movements were based on the change between the initial and final monitoring points. Although trends of generally consistent out-of-slope movement are apparent, no strong correlation between horizontal movement and slope angle could be drawn.

Horizontal movements recorded in the north central portion of the site are typically smaller than movements occurring along other cross-sections with similar surface geometry. This area corresponds to the oldest build-out geometry of the landfill and is the most complex. The relatively small movements recorded in this area may be attributed to two factors. The first factor may be due to the relative stiffness of the older to the younger waste. The second factor may be due to three-dimensional stresses within the landfill mass resulting from the complex geometry.

Three surface monuments were located on the top deck in areas that were thought not to be impacted by slope movement processes. Little or no horizontal movement was recorded for the surface monument on the eastern side of the landfill. Horizontal movements were recorded for the two top deck surface monuments located on the western portion of the landfill and ranged from 0.8 to 1.2 feet. Based on the current understanding of the landfill's bottom geometry (EPA, 1988), these movements may be influenced by the bottom geometry of the landfill.

Vertical Surface Movements: The vertical component of surface movement ranged from approximately 1 to 14 feet. Shown in parenthesis on Figure 2, the annualized rates of settlement ranged from 0.12 to 1.59 feet per year. As observed in Figure 2, most of the data points are located on slope areas. Movements of the slope areas are complex and are influenced by many factors; however, generally, lesser amounts of settlement have been recorded within the older waste portions of the landfill. Settlements recorded at the surface monuments located on the top deck were the greatest and ranged from 7.89 to 13.98 feet (1.15 to 1.59 feet per year). Generally, plots of vertical surface movements versus time appear to be fairly linear, which suggests that movement is occurring at a constant rate.

The recording of 13.98 feet was made at surface monument SM-36. At this location, approximately 289,000 cubic yards of clean soil fill was stockpiled between October 1991 and October 1992 for later use in the final cover system. It should be noted that this location is underlain by the youngest waste. Two stockpiles were sited on the west end of the landfill. The footprint of the south stockpile measured approximately 600 feet by 400 feet, while the north stockpile measured 650 feet by 400 feet. In October 1991, during stock piling activities, SM-36 was buried and no measurements were made. SM-36 is located approximately 20 feet beneath the north stockpile and was excavated in October 1993. Prior to stockpiling activities, a settlement rate of 1.59 feet per year was recorded at SM-36. Major stockpiling activities occurred between October 1991 and October 1992. Measurements made after the excavation of SM-36 indicate 6.95 feet of settlement occurred during the period of July 1991 through October 1993. Recently, a settlement rate of 1.35 feet per year was recorded at SM-36. Assuming a conservative settlement rate of 1.35 feet per year, about 4 feet of settlement could be attributed to placement of the soil stockpile.

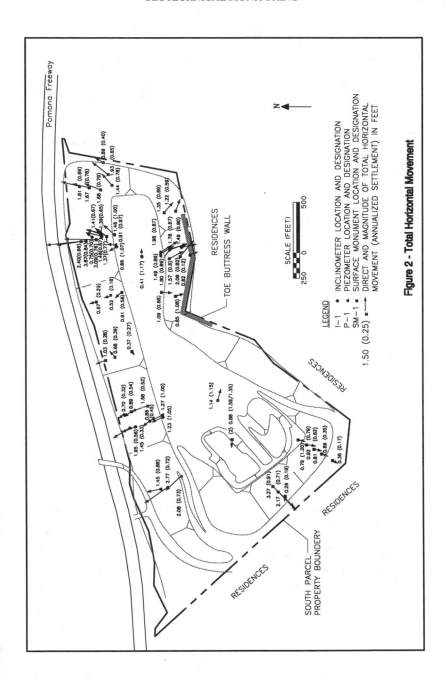

Figure 2 - Total Horizontal Movement

Resultant Vectors of Surface Movements: The horizontal and vertical components of movement were combined and a straight line was drawn from the first point to the last point to determine the resultant vector of movement. This information was then plotted onto 14 cross-sections (EPA 1995a, 1988). Figure 3 presents an example of one of these 14 cross-sections. The resultant vectors of movement ranged from 0.60 to 13.99 feet. Generally, all movements tend to be down and out-of-slope which is characteristic of volumetric and deviatoric creep.

Larger movements were recorded at midslope above the toe buttress wall located on the south eastern side of the landfill. The rigidity of the wall may be influencing these movements. The movements recorded in the north central portion of the landfill are the smallest and may be due to the relative stiffness of the older waste to younger waste and the complex build-out geometries. Cross-Section G, presented in Figure 3, shows that the movement recorded at SM-36 is mostly due to settlement. Also, the build-out geometry in this location is complex because it includes younger waste underlain by steeply sloping older waste.

Subsurface Movements

Six inclinometers were installed with Sondex (corrugated polyethylene tubing) sleeves; four were installed with telescoping sections; and one standard inclinometer was installed. The Sondex and telescoping inclinometers served two purposes. The first purpose was to perform as an extensometer and monitor subsurface settlement. The second purpose was to mitigate downdrag forces and avoid buckling that might otherwise develop from landfill settlement. The standard inclinometer had no provisions to meet either purpose. Provided in EPA 1989 is detailed as-built information regarding the installations.

Lateral Subsurface Movements: To account for the inclinometer installations being "free-floating" within the waste prism, the lateral deflection was determined by the following 2-step procedure. First, using DigiPro™, a computer software package developed by SINCO in 1994, the lateral deflection of the casing was computed by assuming that the bottom of the casing is fixed in translation and rotation. Second, the computed deflection curve is then coupled with positional data by assuming that the top of casing corresponded to the center of the traffic box and translating the deflection curves to match the positional data. The digitilt data was also corrected for orientation, settlement, and casing twist.

Figure 4, shows the resulting inclinometer displacement profiles for inclinometer I-2. The plot shows a general uniform out-of-slope movement. Plots for inclinometers I-1 and I-3 also show a general uniform out-of-slope movement of about 15 to 25 inches. Inspection of the deflection curves suggests that deviatoric creep has occurred for I-1 and I-2 to a depth of 80 feet, and for I-3 to a depth of about 60 feet. Typically, inclinometer deflection curves are inspected to identify an "active-zone" — a zone within the slope in which a potential slip surface may be developing. The deflection curves for I-1, I-2, and I-3 did not show this behavior.

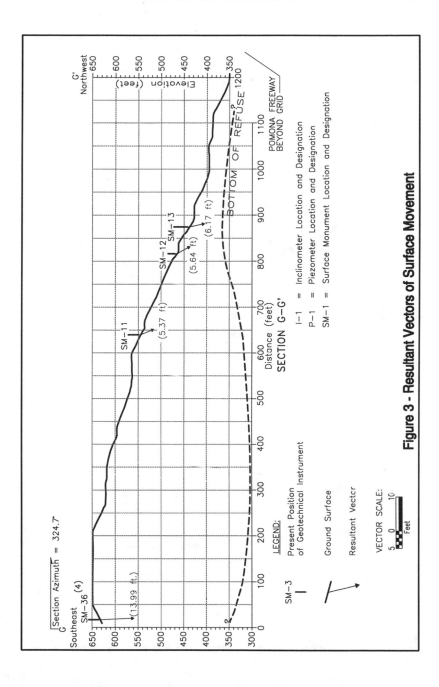

Figure 3 - Resultant Vectors of Surface Movement

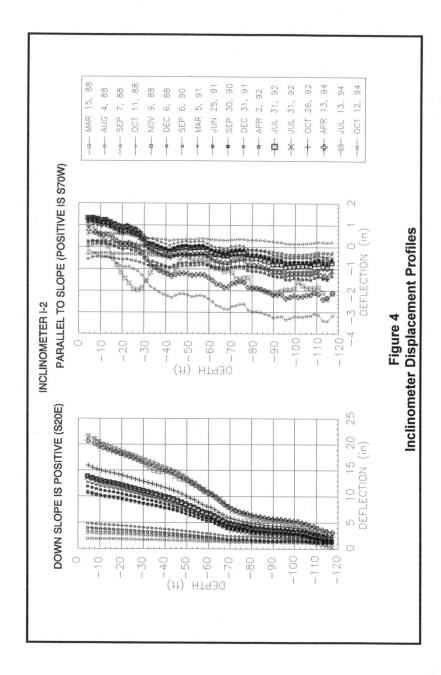

Figure 4
Inclinometer Displacement Profiles

In October 1992, the digitilt sensor in I-3 could not be lowered beyond a depth of 64 feet. In April 1993, total depth of I-3 was reduced to 30 feet. These obstructions could indicate buckling and/or shearing of the inclinometer casing. Inclinometer I-3 is located closest to the toe buttress wall and in an area of steep interior side slope geometry. The rigidity of the toe buttress wall in proximity to the steep interior side slopes may have induced significant stresses within the landfill mass causing the distortions observed in I-3.

Plots of the displacement profiles for inclinometers I-4, I-5, and I-6 generally indicate out-of-slope movement of about 15 to 20 inches. With the exception of I-6, the deflection curves are similar to the deflection curves for I-1, I-2, and I-3. Inspection of the deflection curve for I-6 indicates that an active-zone has developed at a depth of about 10 to 15 feet which suggests that the upper 10 to 15 feet is moving out-of-slope as a block. At this location, depth of cover was reported to be 10 feet during installation of the inclinometer (EPA, 1989). Based on the 1989 cover depth observation, the movement may be occurring at the interface of cover and refuse material at this location. Below this 10 to 15 foot depth, data from I-6 show general out of slope movement as observed for inclinometers I-1, I-2 and I-3.

Inclinometer I-5 collapsed approximately 6 feet below the ground surface between the March and July 1988 monitoring periods. The casing was sheared horizontally; however, no surface formation was visible on the down-slope face. Collapse of the inclinometer could be the result of an isolated subsurface movement which may have resulted from damage to a stormwater drain located in the vicinity of I-5 (EPA, 1989).

Generally, the movements shown by the deflection curves for I-7 through I-11 are different from those observed for I-1 through I-6. Inclinometer I-11 is located within 400 feet of the northeast corner of the landfill and shows subsurface lateral movement of about 16 inches east and 7 inches down slope (i.e., out-of-slope). Note that the displacements in I-11 are mostly in the east direction, which is perpendicular to the down slope direction. A similar trend of movement is observed in I-10, but with more down-slope displacement. Note that the movement of the bottom of inclinometer I-10 is significantly larger (on the order of 10 inches) than any of the other inclinometers. Inspection of the deflection curves for I-9 show a similarity to those for I-6 and suggest that the upper 20 feet is moving out-of-slope as a block. However, it is reported (EPA 1989) that the cover is only 3 feet thick in this area.

The movements shown by the deflection curves for I-10 and I-11 may be impacted by the build-out geometry of the eastern portion of the landfill. Between 1960 to 1964, a large flat-bottomed pit with steep interior slide slopes in excess of 100 feet was excavated into bedrock on the eastern side of the landfill. During the filling of this area between 1964 and 1973, the refuse was placed to match the height of the existing older portion of the landfill to the west. The inclinometer deflection curves for I-10 and I-11 and surveying results of the eastern part of the landfill suggest a block movement of the more recent deposits toward the east. The nature of this movement may be due to a combination of the compressibility of the younger refuse and three-dimensional effects resulting from the complex bottom and interior side slope build-out geometry.

Vertical Subsurface Movements: The sleeved (Sondex) and telescoping inclinometers serve as extensometers in which subsurface settlement is monitored. Settlements are generally larger at the surface than at depth. Figure 5 presents a plot of change in elevation versus time for Sondex inclinometer I-10. Approximately 4.57 feet of settlement occurred at the ground surface at I-10 shown in Figure 5; 1.4 feet occurred at the inclinometer's base. About 90 feet of waste is estimated to underlie the base of this inclinometer.

Visual inspection of the plots of change in elevation, versus time for both the Sondex and telescoping extensometers, showed that the rate of settlement at each depth were generally linear —which suggests that movement is occurring at a constant rate. It was also observed that the rate of settlement was initially greater in the upper 50 feet of each inclinometer installation and appeared to approach a more constant rate after the first one to two years. This observation may be a result of settlement of the grout backfill used in each inclinometer. In I-4, a settlement of 12 feet was recorded in the upper Sondex ring (Ring 1) compared to a ground settlement of about 7.5 feet. The upper 20 feet of the cement bentonite grout in I-4 deteriorated and was reestablished over time using fine sand.

Liquid Levels

Liquid levels were measured in five piezometers that were numbered to match the number of the adjacent inclinometer. The purpose of monitoring porewater is to gain an understanding of landfill mass strength by estimating effective shear strength. The piezometers were installed in saturated zones encountered during drilling. Except for piezometer P-4 (which remained essentially constant), in general, the liquid levels in the piezometers have decreased from 5 to 25 feet since the piezometer installations in 1987.

The greatest fluid decrease occurred in P-7 and P-8. Settlement measured at P-7 was 6.62 feet. The top of the 20 foot long screen was set at elevation 595.80 feet and the December 1987 liquid level was measured at elevation 612, some 16 feet above the top of the screen. The liquid level in P-7 dropped about 15 feet during the first six months of monitoring. After this period, the rate of decrease became fairly constant. P-4 has exhibited a fairly constant liquid level of approximately 546±4 feet.

Deep leachate extraction began on the western side of the landfill in February 1994; P-4 is the only piezometer located in this area. To date, these site activities don't appear to have affected the liquid levels in P-4. Also, due to seasonal variations in precipitation, no impacts to the trend of liquid levels in the five piezometers have been observed. It should be noted, however, that since the monitoring period began — with the exception of February 1993 — the site has experienced below average precipitation.

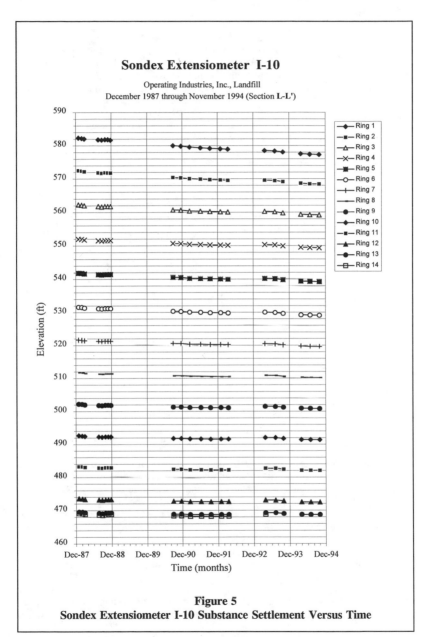

Figure 5
Sondex Extensiometer I-10 Substance Settlement Versus Time

Temperature

During the period from December 1990 through July 1994, temperature was not recorded. Temperature measurement was reinitiated in July 1994 to develop a data base, which may be used to evaluate deviatoric creep. Table 1 presents a summary of temperature data obtained in inclinometer I-10. Temperatures within the inclinometers are variable over time, but the data demonstrate a temperature increase with depth. Although difficult to determine trends because of gaps in the data base, particularly from 1990 through 1994, the data indicates a slight increase in temperature over time in I-3, I-4, and, I-7 and to a lesser degree, in I-10 and I-11. Also, in inclinometer I-6 the temperature increased to a depth of about 60 feet then it decreased with depth. This decrease may be attributed to moving from a zone of young waste to a zone of the oldest waste.

TABLE 1

TEMPERATURES INSIDE SLOPE INCLINOMETER I-10
FROM 12/22/87 TO 10/13/94

DEPTH (ft)	12/22/87	3/14/88	7/1/88	10/10/88	12/6/88	9/10/90	7/12/94	10/13/94
Ambient	49.0	74.5	98.0	80.0	*	102.0	87.4	86.0
Top of Casing	61.8	72.5	98.0	85.0	*	-	-	86.0
10	81.3	80.0	98.0	90.0	*	100.0	89.1	97.0
20	90.8	93.5	101.5	100.0	*	105.0	104.6	109.0
30	96.0	98.5	105.5	101.0	104.0	114.0	117.9	119.0
40	97.5	96.0	104.5	107.0	109.0	119.0	124.7	123.0
50	94.0	97.5	106.0	109.0	112.0	123.0	128.2	127.0
60	95.0	98.5	107.5	113.0	115.0	125.0	132.5	131.0
70	95.8	100.5	108.5	117.0	118.0	130.0	138.4	136.0
80	96.3	101.0	109.5	119.0	119.0	132.0	139.3	138.0
90	99.8	102.5	110.5	123.0	122.0	133.0	141.0	141.0
100	101.5	107.5	114.5	130.0	130.0	135.0	141.7	142.0
110	104.8	110.0	116.0	131.0	132.0	137.0	-	140.0
Bottom of Casing	107.5	111.0	117.5	134.0	133.0	-	-	140.0

NOTES:
1. "-" Data not recorded at this location
2. "*" Data not used in analysis

DISCUSSION

The purpose of collecting geotechnical monitoring data over 30 periods of monitoring at the OII Landfill (from December 1987 through October 1994) was to help assess overall slope movements, distinguish between cover and refuse mass movements, and evaluate time-dependent deformations. Based on the results of this study, the OII Landfill slope movement appears to be influenced by multiple factors including, but not limited to, the heterogeneity, age, and relative compressibility of the waste, build-out geometries underlying the slopes, bottom and interior side slope geometry, and the physical and chemical nature of the waste. To further evaluate the basis of these movements, a fully integrated approach that incorporates a synthesis of the existing data should be applied.

The assessment of landfill slope movements rely heavily on surveying results. Based on the results of a detailed geodetic deformation analysis (EPA, 1995b), it was shown that statistically significant displacements have occurred at the OII Landfill. It was noted that for any monitoring installation, a data quality objective process should be included with a view to enhancing the cost-effectiveness and reliability of the survey results and their value for the geotechnical interpretation (in surveying terms, this process is termed pre-analysis). This can be accomplished by establishing a stable network (i.e., monuments grounded into bedrock and sleeved, if necessary), and development of a set of standards, specifications, and procedures that result in the measurement accuracy and precision desired.

CONCLUSIONS

Based on the information presented in this paper, the following general conclusions can be made:

1.   The general trend of movement for the OII Landfill slopes is down and out-of-slope. These movements have occurred at a fairly constant rate. Generally, the displacement vectors of movement show the landfill is, in effect, shrinking.

2.   Observed movements of the eastern portion of the OII Landfill (Figure 2) may be attributed to a combination of the build-out geometry and construction of the toe buttress wall. Additionally, the movements in Inclinometers I-10 and I-11 suggest that the northeast corner of the OII Landfill may be experiencing translational block movement; possibly along an intermediate cover interface within the refuse prism.

3.   The rate of settlement in SM-36, which was about 1.56 feet per year prior to placement of the clay stockpile, appears to be stabilizing at about 1.35 feet per year. If a settlement rate of 1.35 feet per year is assumed, an immediate settlement of approximately 4 feet can be attributed to the placement of the clay stock pile.

4.   The deflection curves shown for Inclinometers I-6 and I-9 indicate that a slip surface has developed at a depth of approximately 10 feet and 20 feet, respectively. These are the only inclinometers in which mass movement between cover and refuse can be discerned.

5.   The horizontal and vertical movements recorded in the north central portion of the landfill suggest that the older refuse materials and the unusual build-out geometry significantly impact movement.

ACKNOWLEDGMENTS

The data compilation and interpretation work discussed in this paper was performed by CDM Federal, under contract to the U.S. Army Corps of Engineers, for EPA. CDM Federal is providing oversight services to EPA, Region IX, the lead agency for the OII Landfill project. Although the work was funded by the EPA, the preparation of this paper was not funded by and has not been subject to the Agency's review. Therefore, this paper does not necessarily reflect the Agency's views, and no official endorsement should be inferred.

The authors wish to acknowledge the contribution of colleagues Bruce Chaplin, Bill Orr, and Dr. John Zellmer who were responsible for much of the planning and execution of the work. In addition, this work would not be possible without the direction received from EPA's Technical Review Panel comprised of Drs. Mary E. Hynes, I.M. Idriss, Ellis L. Krinitzsky, Arvid O. Landva, and Raymond B. Seed. We are also grateful to Keith Grubb and Jeff Di Donato for their excellent word processing and drafting assistance.

REFERENCES

U.S. Environmental Protection Agency. 1988. "Literature Review and Existing Data Review Technical Memorandum Geotechnical Task RI/FS2 South Parcel Operating Industries, Incorporated, Monterey Park, California." Prepared by CH2M Hill. May 13.

U.S. Environmental Protection Agency. 1989. Agency Review Draft Technical Memorandum Data for the Geotechnical Site Assessment RI2/FS, South Parcel, Operating Industries, Incorporated, Monterey Park, California. Volumes I, II, & III. Prepared by CH2M Hill. February 6.

U.S. Environmental Protection Agency. 1994. Draft Remedial Investigation Report for Operating Industries, Inc. Landfill Superfund Site, Monterey Park, California, Volumes I & II. Prepared by CH2M Hill. October 25.

U.S. Environmental Protection Agency. 1995a. Geotechnical Monitoring Data Compilation and Interpretation Report for December 1987 to October 1994, Operation Industries, Inc. Landfill Superfund Site, Monterey Park, California. Prepared by CDM Federal Programs Corporation. February 10.

U.S. Environmental Protection Agency. 1995b. Volumes I, II, and III, Final Report, Subsidence Monitoring of the Operating Industries, Inc. Superfund Site, Monterey Park, California. Prepared by Measurement Science, Inc. February 10.

The Performance of Two Capillary Barriers During Constant Infiltration

John C. Stormont[1]

John C. Stormont[1]

Abstract

Capillary barriers, consisting of fine-over-coarse soil layers, have been suggested as an alternative cover component for waste disposal facilities constructed in dry climates. Water is held in the fine layer by capillary forces until the water is removed by evaporation, transpiration or percolation. If the fine-coarse interface is sloped, water in the fine layer can also be diverted laterally along the fine-coarse interface. To increase the lateral diversion capacity, layered capillary barriers are being considered which have an increased unsaturated hydraulic conductivity of the fine layer in the downdip direction. Homogeneous and layered capillary barriers have been installed and tested to compare their performance. Water balance measurements were made in above-grade "boxes" consisting of a 90 cm thick fine layer over a 25 cm coarse gravel layer. The boxes were 7 m long and built on a 5% grade, and were covered to limit evaporation and transpiration. One box had a homogeneous fine layer, while the fine layer in the other box was layered to increase its ability to laterally divert water. The barriers were stressed by adding water to the top of each profile for 74 consecutive days at a rate of about 0.5 cm/day. Measurements were made of water storage and lateral diversion in the fine layer, and breakthrough into the coarse layer. More than 99% of the collected water had laterally drained from the layered capillary barrier, whereas more then 75% of the collected water from the homogeneous capillary barrier broke through into the coarse layer.

Introduction

Landfills, surface impoundments, waste piles, and some mine tailings are required to be covered with an engineered cover or cap upon closure. There is a need for improved cover designs. Many conventional designs feature a compacted soil layer, which can be difficult and costly to construct, and suffer from desiccation, root and animal intrusion, and other concerns. Geosynthetics, increasingly being used in cover designs, will typically have some construction

---

[1] Sandia National Laboratories, MS 0719, Albuquerque, NM 87185

flaws, introduce planes of weakness with adjacent materials, and have a limited data base regarding their long-term durability. Multicomponent covers combine numerous components to achieve specific functions and satisfy multiple regulations, but can be very expensive and have a doubtful prognosis for meeting long-term design objectives (Daniel, 1994).

Capillary barriers, consisting of fine-over-coarse soil layers, have been suggested as an alternative cover component in dry climates. Water is held in the fine layer by capillary forces until the water is removed by evaporation, transpiration or percolation. Water in the fine layer can also be diverted laterally near the fine-coarse interface if the fine-coarse interface is sloped. The relatively simple configuration of a capillary barrier will result in a lower cost than most other systems. A capillary barrier functions because of the contrast in hydraulic conductivity between the fine and coarse soils at similar suction heads which exist near the fine-coarse interface. The functional performance of a capillary barrier can be explained by considering Figure 1. Beginning at relatively dry conditions, that is, at high soil suctions, the fine soil has a finite hydraulic conductivity, whereas the hydraulic conductivity of the coarse layer will be immeasurably small. With increasing water content and decreasing suction head, the hydraulic conductivity of the fine layer will increase gradually. The hydraulic conductivity of the coarse layer will remain immeasurably small until its water entry head is overcome. Under these conditions, water will not move from the fine layer into the coarse layer, but will instead increase the water content of the fine layer and/or be diverted laterally in

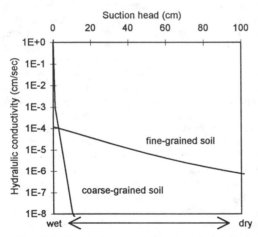

Figure 1 - Typical hydraulic conductivity of fine- and coarse-grained soils for a capillary barrier.

the fine layer if the fine-coarse contact is sloped. Laterally diverted water will result in increasing water content in the downdip direction. The diversion length is the distance along the fine-coarse interface which water is diverted before there is appreciable breakthrough. Breakthrough into the coarse layer occurs when the suction head at the contact equals the water entry head of the coarse layer. When the suction head decreases below this value, the hydraulic conductivity of the coarse layer will increase rapidly and eventually exceed that of the fine layer. Under these conditions, the fine-over-coarse arrangement is not a barrier to downward water movement.

A capillary barrier remains effective if the combined effects of evaporation, transpiration, and lateral diversion equals or exceeds the infiltration from precipitation events, thereby keeping the fine layer sufficiently dry so that appreciable breakthrough into the coarse layer does not occur. Depending on the climate and thickness of the fine layer, evaporation and transpiration may remove sufficient water to prevent failure, and the capillary barrier need not be sloped for lateral diversion. Wing (1993) reported that non-sloping capillary barriers, comprised of a 1.5 m thick soil layer over a coarse layer, were capable of storing and removing three times the average annual precipitation near Hanford, WA (52 cm). At Idaho Falls, a 1.6 m thick soil layer was capable of storing and removing 37 cm of precipitation, which corresponds to the maximum annual precipitation over a 40 year period (Anderson et al., 1993).

In some instances, however, it may be necessary or desirable to account for the diversion capacity of the capillary barrier. In climates which receive significant amounts of precipitation during periods when plants are not active, near-surface soils can approach saturation and capillary barriers become susceptible to failure. Accounting for lateral diversion may reduce the thickness of the fine layer necessary for storage, especially considering unusual or extreme climatic conditions. Lateral diversion can also be considered as a redundant component in a cover system.

The lateral diversion capacity of sloped capillary barriers have been measured in two recent field tests. Hakonson et al. (1994) evaluated the performance of cover systems including two capillary barriers at a site near Ogden, Utah from January 1990 to October 1993. Both capillary barriers consisted of 150 cm of vegetated soil overlying 30 cm of gravel. Both the surface and the soil-gravel interface were sloped at four percent and had a 10 m length in the direction of the slope. Percolation (breakthrough into the coarse layer) was found to occur principally during the late winter and spring (February to May). Nyhan et al. (1990) reported on the water balance of cover systems incorporating capillary barriers over a 3 year period at Los Alamos, New Mexico. The capillary barriers consisted of 71 cm of a vegetated sandy loam topsoil overlying 46 cm of gravel. The fine-coarse interface was sloped at five percent over a 3 m length. One of the two capillary barriers produced no measurable lateral diversion or percolation, attributable to enhanced evapotranspiration of the vegetation on this cover. Percolation and lateral drainage out of the other barrier occurred during snowmelt in two of the three winters.

Measurable percolation in the above cited field tests indicates these capillary barriers did not laterally divert sufficient water to prevent breakthrough over their relatively short lengths (10 and 3 m). These short diversion lengths are a consequence of the relatively low hydraulic conductivity of the fine-grained soils compared to the infiltration rate during stressful periods. Even if a soil with a higher hydraulic conductivity is readily available, it may not possess the moisture retention characteristics necessary for the top layer of a capillary barrier (Goode, 1986). An alternative means to increase the diversion length of a capillary barrier is to incorporate hydraulic conductivity anisotropy in the fine layer (Stormont, 1995). If the fine layer has a greater conductivity in the lateral direction, its ability to divert water laterally will be improved. In a homogenous material, hydraulic conductivity anisotropy due to particle orientation can be induced by compaction, but is generally limited to less than one order of magnitude. A heterogeneous profile created by layering can be considered as effectively anisotropic, and differences in directional hydraulic conductivity of many orders of magnitude are possible.

In this study, the capacity of two unvegetated capillary barrier designs to divert a constant infiltration rate was measured. One barrier utilized a homogeneous profile, while the other was layered to create a profile with an increased lateral diversion capacity. The principal objective of the study was to gain insight into means to increase the diversion capacity of capillary barriers.

Test Configuration

The capillary barriers were constructed and tested in two above-grade wooden structures (boxes) configured to measure the water balance (Figure 2). The boxes were 7.0 m long, 2.0 m wide and 1.2 m high, and were sloped at 5%. The bottom and sides of the boxes were covered with 10 mil polyethylene sheeting. Gravel was placed in the bottom 25 cm of the box from the updip wall to within 1.0 m of the downdip wall of the box. A fine-grained material - local soil in one box, and layers of local soil and sand in the other box - was placed on top of the gravel to the top of the box. A region for laterally diverted water to accumulate and drain was created in the last 1.0 m of the box by completely filling this region with the fine-grained material. At 30 cm and 60 cm above the fine-coarse contact, aluminum drip edging was attached to the edge of the box to reduce the potential for preferential flow along the wall. A 170 g/m^2 nonwoven geotextile was placed between the two materials to prevent migration of fines into the coarse layer.

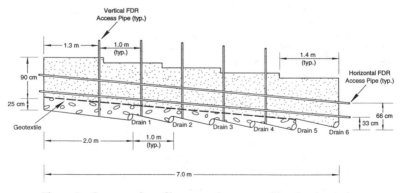

Figure 2 - Cross-section of boxes used to test capillary barrier designs.

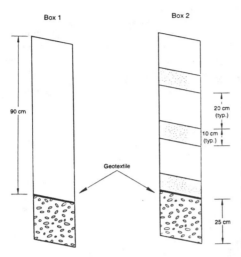

Figure 3 - Profiles of capillary barrier designs.

The homogeneous design utilized local soil as the 90 cm thick fine-grained layer (Figure 3), and was designated Box 1. The layered design, designated Box 2, consisted of alternating 10 cm thick layers of a sand and 20 cm thick layers of local soil. The first layer on top of the geotextile was sand, followed by a soil layer, then sand and so on. The fine layers were compacted with a hand-held pneumatic tamper to yield an in place density of about 1.65 g/cc. The combined effect of water added during compaction with the antecedent moisture in the soil yielded an initial water content of between 12% and 15%

The soil at the top of the boxes was terraced into five, 1.4 m long, horizontal intervals to prevent run-off when adding water. Water was added by means of a gravity-fed drip irrigation system. Sixty five liters (0.48 cm) were added daily for 74 consecutive days from May 23 through August 4 at a rate slow enough to just prevent ponding. The infiltration rate is equivalent to a flux rate of 1.8×10^{-5} cm/sec. The top of the box was covered with fiber-reinforced plastic to minimize evaporation of water, discourage plant growth, and prevent rainfall from contacting the soil.

Five cm diameter perforated PVC pipes within the boxes served as drains. Drains were spaced along the bottom of the coarse layer and discretized the coarse layer into five collection regions (Drains 1 to 5) providing a means of identifying the breakthrough location. A drain (Drain 6) wrapped in geotextile was also located within the downdip collection region of the fine layer to collect laterally diverted water. Because water was infiltrated above the lateral diversion collection area, some of the water produced by this drain was vertical infiltration and not lateral diversion. All of the drains were sloped at an angle of 2 degrees from the horizontal, and emptied into below-grade sumps for the measurement of the collected water.

Soil moisture changes were measured with a frequency-domain reflectometry (FDR) probe. This non-nuclear device determines soil water content by measuring the capacitance of the soil, and operates from vertical and horizontal 5-cm diameter PVC access tubes (Figure 2). The FDR probe was calibrated prior to its use in the field using soil and sand from the field site at a known moisture content and density.

Although the vertical access probes penetrated the gravel layer, the FDR device was ineffective at measuring the small amount of water typically found in the gravel layer. Moisture content measurements were also made with time-domain reflectometry (TDR) probes embedded in the fine layer at various locations along the length of the box. Data from the TDR probes were compromised by power outages, temperature effects, and software problems, and are not reported here.

Material Properties

The soil used to construct the fine layers of the boxes was a silty sand obtained from within one meter of the surface. This soil has a median particle diameter of 0.11 mm, and 30% of the soil is finer than 0.075 mm. The soil has a porosity of about 40%, and a saturated hydraulic conductivity of 1.2×10^{-4} cm/sec. The layered design incorporated a sand into the overlying fine layer. The sand is uniform, with 99% of the sand between 0.6 and 0.08 mm and a mean particle diameter of 0.3 mm. The sand has a porosity of about 40%, and a saturated hydraulic conductivity of 2.1×10^{-2} cm/sec. The sand and soil particle size distributions satisfy a conventional filtering criterion to limit migration of one into the other. A poorly-graded rounded stone was used as the coarse layer material. The gravel has a mean particle diameter of about 2 cm. The saturated hydraulic conductivity was estimated to be 10 cm/sec by the method of Sherard et al. (1984).

Van Genuchten's (1980) functions were used to describe the volumetric moisture retention and unsaturated hydraulic conductivity of the materials used in these tests. The volumetric water content θ and pressure head h are related by means of

$$\Theta = \frac{\theta - \theta_r}{\theta_s - \theta_r} = \left[1 + (\alpha h)^n\right]^{-m} \qquad (1)$$

where Θ is the dimensionless water content, the subscripts s and r indicate the saturated and residual values of the water content, α, m and n are fitting parameters, and $m=1-1/n$. The hydraulic conductivity is given as

$$K = K_{sat} \, \Theta^{1/2} (1 - (1 - \Theta^{1/m})^m)^2 \qquad (2)$$

The hydrologic properties are given in Table 1. The properties for the sand and soil were derived from standard laboratory tests. The van Genuchten parameters of the gravel were taken from Fayer et al. (1992).

Table 1 - Hydrologic properties for materials used in experiments.

	K_{sat} (cm/sec)	α (cm^{-1})	n	θ_r	θ_s
soil	1.2×10^{-4}	.021	1.87	0.08	0.40
sand	2.1×10^{-2}	.039	5.74	0.06	0.39
gravel	10	4.93	2.19	0.005	0.42

The unsaturated hydraulic conductivities as a function of suction head for the sand, soil and gravel are given in Figure 4. For the system of alternating soil-sand layers, the sand layers can serve as either a capillary barrier or as a "transport" layer depending on the value of the suction head at the interface with an overlying soil layer.

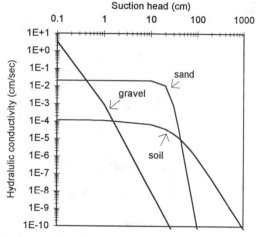

Above about 40 cm, the soil is more conductive than the sand and a capillary barrier is formed. Once the suction head at the interface decreases to about 40 cm, the sand layer no longer impedes downward flow and can accommodate as much water as the overlying soil layer can provide. The sand layer hydraulic conductivity is about two orders of magnitude greater than that for the overlying soil layers at suction heads of 30 cm and

Figure 4 - Hydraulic conductivity vs. suction head for materials used in experiments.

less. The bottom sand layer, underlain by the gravel, will retain water until the suction head decreases to about 1 cm (the water entry head of the gravel) when breakthrough into the gravel will occur.

Breakthrough and lateral drainage

The breakthrough into the coarse layer and the water laterally diverted and drained is given on Figure 5 for Box 1. The lateral diversion drain (drain 6) was the first to produce percolate 43 days after infiltration began. Breakthrough into the coarse layer first appeared in drain 3 on day 47, followed by drain 2 and 4 (day 50), drain 5 (day 53), and drain 1 (day 73).

Wet soil and plant growth where the drain pipes exited the box indicated that the drains were leaking and not all of the water was being collected. Drains 1 through 5 were repaired on day 53, and an increase in percolate collection was noted. Modification to drain 6 on day 65 increased percolate production, but there was still evidence of loss of some water from the lateral diversion drain region. The water collection rate of the coarse layer drains 2 through 5 continued to increase with time, all at a similar rate of about 0.25 liter/day. Drain 1 produced water near the end of the test. Once the lateral diversion drain was reworked, the water collection rate remained relatively constant at about 7 liter/day.

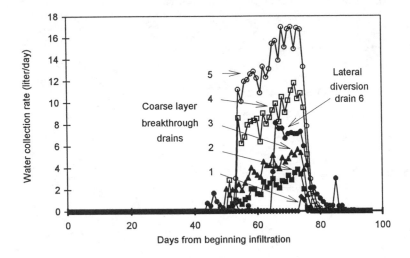

Figure 5 - Water collected from Box 1.

The amount of water collected from Box 1 on the last day of infiltration (day 74) is given in Figure 6a. The greatest amount of breakthrough increased with downdip distance, with more than 30% of the collected water being collected in drain 5. About 17% of the collected water was from the lateral diversion drain, which is about the amount infiltrated in the interval above this drain. The water collected on the final day from both coarse layer breakthrough and fine layer diversion was about 45 liters, which is about 70% of the daily infiltration rate. Once infiltration ceased, the percolate production and lateral drainage out of Box 1 dropped quickly. Within 1 week, breakthrough into the coarse layer had ceased. Lateral drainage dropped to zero 18 days after infiltration had stopped.

The water collected from Box 2 drains is given in Figure 7. Almost all of the water which exited this box was collected in drain 6, that is, it drained laterally out of the fine layer and did not breakthrough into the coarse layer. Water was first collected from drain 6 on day 35, and increased rapidly to about 35 liter/day by day 45. The water collection rate continued to increase by about 0.5 liters/day until infiltration ceased. About 1 liter total was collected from drain 5 and none from drains 1 through 4. There was some evidence of possible leakage from drain 6, but not from any of the other drains.

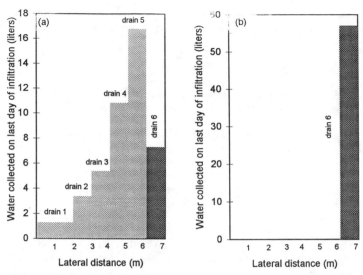

Figure 6 - Water collected in (a) Box 1 and (b) Box 2 on the last day of infiltration.

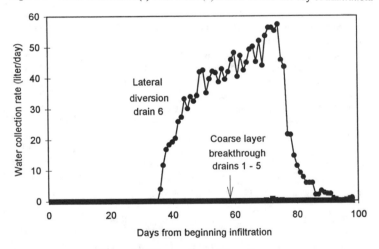

Figure 7 - Water collected from Box 2.

The amount of water collected from Box 2 on the last day of infiltration is shown in Figure 6b. All of the collected water was from drain 6, that is, laterally diverted to the fine-layer drain area. No water was collected from drains 1 through 5 on this day. The lateral drainage was about 87% of the daily infiltration rate. Once infiltration ceased, water production from drain 6 continued at a decreasing rate. Small amounts of water (less than 1 liter/day) were being collected 20 days after infiltration ceased.

Water content measurements

There were some difficulties with the water content measurements using the FDR probe. The moisture content measurements at the first cable stop near the top of the boxes indicated an essentially zero water content. The probable cause was that the access pipes had been rocked back and forth - by wind or from movement of the plastic cover - to cause a slight annulus around the access pipe. This space would significantly impact the resulting water content measurement. This effect is being investigated for influencing (lowering) measured values deeper in the profile as well. Another difficulty was in interpreting the measured values in the layered profile of Box 2. The manufacturer's specifications for the FDR indicated it sensed in a disk-shaped region approximately 10 cm radially into the soil and 5 cm parallel to the probe, and cable stops were selected to coincide with the center of each sand and soil layer. Water content measurements on samples collected from the box profile indicated that the FDR measurements were influenced by water contents of adjacent layers. A separate calibration for the measurements in the layered profile was developed using the water contents from the samples.

Water contents determined from the FDR measurements are shown in Figure 8 for the vertical access pipes at the mid-length of Box 1 and Box 2. These values are comparable to values from other vertical and horizontal access pipes, with only a small increase in water contents near the downdip edge of the boxes. Water contents are given for a number of times, including the initial condition prior to the addition of water, after 24, 44 and 74 days of infiltration, as well as 21 days after infiltration stopped (95 days from the beginning of infiltration). No meaningful water content measurements were made in the coarse layer because of the large voids in the gravel. The gravel layer was assumed to be dry, or if water had broken through, it was assumed to drain instantaneously. In the figures, the Box 1 data were connected with straight lines to imply relatively smooth changes in water content in the homogeneous layer. The Box 2 data were not connected with lines because of the layered structure; a constant value in each layer may be a better assumption.

The wetting front can be seen in both boxes after 24 days of infiltration, reaching the fine-coarse interface in Box 1 and the last soil layer in Box 2. There had not yet been lateral drainage or breakthrough into the coarse layer for either box. After 44 days, water had been laterally diverted in Box 2 for about 10 days and Box 1 first produced lateral drainage. In Box 1, water contents near the fine-coarse interface were near the saturation of the soil, consistent with the impending breakthrough of water into the coarse layer. Within 3 days of this measurement, breakthrough into the coarse layer of Box 1 was detected. The soil in the lower portions of Box 2 were appreciably drier than comparable depth in Box 1, and the bottom sand layer had a water content corresponding to a saturation of about 75%.

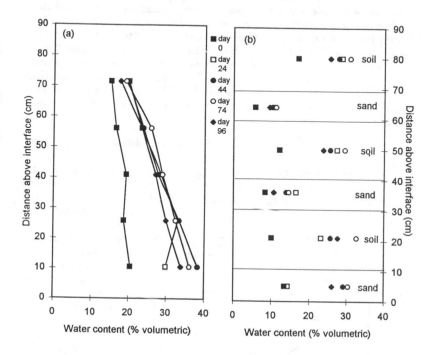

Figure 8 - Water content measurements in (a) Box 1 and (b) Box 2 from vertical access pipes at mid-length of boxes.

On the last day of infiltration (day 74), the water contents in Box 1 were reduced somewhat over those on Day 44 because of the breakthrough of water into the coarse layer over the length of the box. Once water moved into the coarse layer, the water entry head of the coarse layer was overcome and a greater suction head and lower water content resulted in the fine layer near the interface. The water contents near the interface in Box 2 increased somewhat over this same period of time, consistent with the capillary barrier remaining intact. For both boxes, 21 days after infiltration ceased, the water content decreased at all measurement locations as water continued to drain from the fine layer and perhaps evaporate from the top of the profile. The decreases in Box 2 are larger than Box 1, consistent with the larger post-infiltration drainage from Box 2.

Water balance

The general water balance for this experiment consists of infiltration (added water), surface run-off, changes in soil water storage in the fine layer, changes in soil water storage in the coarse layer, lateral drainage, drainage out of the coarse layer (percolation), and evapotranspiration. Because the experiment was designed to quantify lateral drainage and percolation, some components of the water balance were eliminated or minimized. The fine layer was terraced so that there was no run-off when water was added. Because the boxes were covered, plant transpiration was precluded and evaporation was minimized. The large stone used as the coarse layer material has a field capacity of <1%, so the storage capacity in this layer was assumed to be zero. Incorporating these simplifications, the water balance equation reduces to the values measured in the experiment.

$$infiltration - \Delta\ fine\ layer\ water - lateral\ drainage - percolation = 0 \qquad \cdot (3)$$

The water balance for both boxes is shown in Figure 9, and reveals that the water balance equation using measured values does not close. Causes for the discrepancies include experimental problems such as leakage from the drainage collection system and uncertainties with water content measurements. Another factor affecting the water balance of the box system is the accumulation of water in the fine layer drain area (6 to 7 m). This region becomes very wet as laterally diverted water accumulates and eventually becomes saturated. Water content measurements were not made in the region immediately surrounding the drain and thus did not indicate the saturated region.

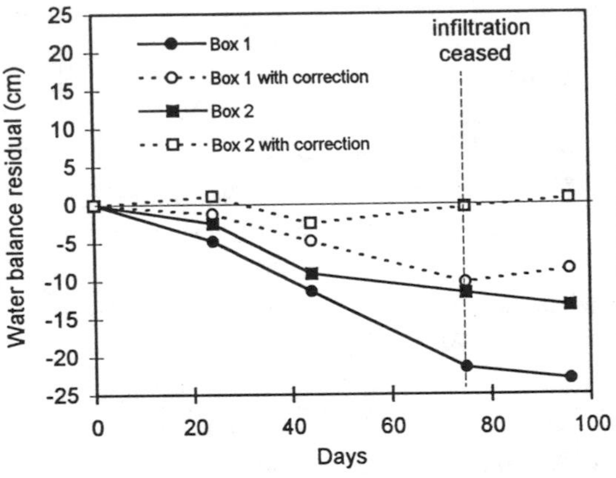

Figure 9 - Water balance residual with and without term for evaporation

Another cause for the water balance discrepancy may be not accounting for evaporation. Water was added at a rate just less than that which would cause ponding, so the near-surface was quite wet. Because the cover over the box was not air-tight and the water was usually added in the late morning hours of warm summer months, water in near-surface would be vulnerable to evaporation. Evaporation can be estimated by assuming the measured water balance terms (the left hand side of Equation (3)) sum to a term accounting for evaporation rather than zero. Consider the water balance data for the first 23 days during which water was added. During this period, there is no breakthrough into the coarse layer or collected lateral drainage in either box, and solving Equation (3) yields an average value of 0.15 cm/day. Although the correction may be attributed to evaporation, the correction term could include other systematic errors in the water balance. The water balance incorporating the evaporation correction term is given in Figure 9. With the correction, the water balance is near zero for Box 2. The water balance for Box 1 is also improved, but because of the leaks in the drainage system of Box 1, closure would not be expected.

Discussion

The experimental results indicate a significant difference in the diversion length for the homogeneous profile (<2 m) and layered profile (>6 m). An expression developed by Ross (1990) and modified by Steenhuis et al. (1991) provides an estimate of the steady-state lateral diversion length L for a capillary barrier subjected to a constant infiltration rate

$$L \leq \tan\phi \left[a^{-1}(\frac{K_{sat}}{q} - 1) + \frac{K_{sat}}{q}(h_a - h_w^*) \right] \qquad (4)$$

where K_s is the saturated hydraulic conductivity, q is the infiltration rate, a is a constant used in the quasi-linear unsaturated hydraulic conductivity formulation, h_a is the air entry head of the fine layer, h_w^* is the water entry head of the coarse layer (estimated to be about 1 cm), and ϕ is the inclination of the fine-coarse interface. For the soil used in the homogeneous profile of Box 1 ($K_s = 1.2 \times 10^{-4}$ cm/sec, $a^{-1} = 20$ cm , $h_a = 20$ cm), the diversion length is estimated to be on the order of 10 cm. This distance is consistent with the breakthrough measured in the experiment in the drain 1 interval (0 to 2 m).

Although Equation (4) was developed for a homogeneous profile, it may be applicable to the layered profile as well. Ross showed that most of the lateral diversion occurs within a distance of $(a \cos\phi)^{-1}$ of the fine-coarse interface. For the sand layer properties ($K_s = 2.1 \times 10^{-2}$ cm/sec, $a^{-1} = 4$ cm, $h_a = 10$ cm), this distance is less than the 10-cm thickness of the bottom sand layer. Consequently, most of the lateral diversion of the layered design may occur in the bottom sand layer, and applying Equation (4) with the sand properties yields a diversion length of a little more than 6 m. The experimental results are consistent with this prediction in that they indicate the diversion length was in excess of 6 m. Another approach for evaluating the layered profile is to consider it to be an effectively anisotropic medium, and use expressions developed for the diversion length of anisotropic capillary barriers (Stormont, 1995). The diversion length calculated this way is also about 6 m.

In Figure 10, the hydraulic conductivity of the layered profile is given as a function of depth for the last day of infiltration. The hydraulic conductivity was calculated using the measured water contents from the center of each layer and the van Genuchten fitting parameters. Although these values are only approximations,

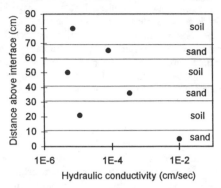

Figure 10 - Calculated hydraulic conductivity vs. depth for Box 2 on last day of infiltration.

they do reveal interesting behavior of the sand and soil layers. Both of the upper sand layers were more conductive than the neighboring soil layers, indicating they are not impeding downward water movement. The bottom sand layer had a measured water content of 30% which corresponds to a suction head of about 20 cm and a hydraulic conductivity of about $1x10^{-2}$ cm/sec. Assuming all the water infiltrating into the profile is passing into this layer, its diversion length is given by $(K \sin\phi \ b)/q$, where b is the layer thickness. The diversion length calculated this way (3 m) is less than the measured diversion length (> 6 m). This result may be attributable to a portion of the sand layer very near the interface having a greater water content and corresponding hydraulic conductivity than the value from the middle of the layer.

The expected performance of a capillary barrier will depend to a large extent on the assumed infiltration rate into the fine layer (e.g., see Equation (4)). When derived from an average yearly precipitation, infiltration rates of 10^{-6} to 10^{-7} cm/sec are reasonable for many sites in dry climates. However, at the shallow burial depth of landfill covers, infiltration from precipitation events will be less attenuated and more episodic than at greater depths. The amount of water in the fine layer varies throughout the year, and the fine layer can become very wet during periods when plant transpiration is low. In this condition, capillary barriers are vulnerable to precipitation moving rapidly through the fine layer and into the coarse layer because the hydraulic conductivity of the fine layer is relatively great and its ability to store additional water is low. A layered design (or a more truly anisotropic design with thinner layers) may have some advantages in this scenario. Depending on the materials used, a layered profile will contain less water for a given infiltration history because it laterally drains water more effectively than a homogeneous design. It would then have more available storage capacity in addition to a greater diversion capacity in response to stressful precipitation events. Another design option may be to incorporate a layer with a low saturated hydraulic conductivity in the upper portions of the profile to reduce the maximum infiltration rate which reaches the underlying transport layers.

Conclusions

The layered design is much more effective in laterally diverting water than the homogeneous design. With a constant infiltration rate of about 0.5 cm/day, all but 1% of the water which moved out of the fine layer was laterally diverted beyond the 6 m length of the capillary barrier. The bottom sand layer served as a "unsaturated transport" layer, and was responsible for most of the lateral diversion of this barrier. The upper two sand layers did not significantly impact the capillary barrier performance at the infiltration rate used in this experiment. At a lower rate, the sand layers would create capillary barriers with the overlying soil layers and impede the downward movement of water.

The very short diversion length of the homogeneous profile (<2 m) is consistent with other tests of capillary barriers, and is a consequence of the relatively low hydraulic conductivity of the soil compared to the infiltration rate. The lower portion of the fine layer increases in water content, with little if any lateral diversion, until breakthrough into the coarse layer occurs.

The steady-state solutions for diversion length by Ross (1990) and others appear consistent with the experimental results, although the rate of water collected from almost every producing drain continues to increase with time. These solutions may be useful if modified to account for the transient infiltration events typical for most near-surface cover systems.

The experiments reported here indicate that capillary barriers can be designed to increase their diversion capacity which, in turn, can reduce the amount of water which moves downward through the barrier. Designing capillary barrier with substantial lateral diversion capacity to complement their ability to store and evapotranspirate soil moisture may increase the useful applications of this alternative barrier technology. Areas of future work include accounting for daily variations in near-surface water balance terms (i.e., non-constant infiltration from varying precipitation and evapotranspiration) in capillary barrier design.

Acknowledgments

This work was supported by the United States Department of Energy Office of Technology Development under Contract DE-AC04-94AL85000. G. Reynolds, J. Kelsey and M. Burkhard provided invaluable assistance in conducting the field measurements.

References

Anderson, J. E., Nowak, R. S., Ratzlaff, T. D. and Markham, O. D. (1993). "Managing soil moisture on waste burials sites in arid regions." *J. Environ. Qual.,* 22, 62-69.

Daniel, D. E. (1994). "Surface Barriers: Problems, Solutions, and Future Needs." *Proceedings of the Thirty-Third Hanford Symposium on Health and the Environment,* edited by G. W. Gee and N. R. Wing, Pasco, WA, 441-487.

Fayer, M. J., Rockhold, M. L. and Campbell, M. D. (1992). "Hydrologic Modeling of Protective Barriers: Comparison of Field Data and Simulation Results." *Soil Sci. Soc. Am. J.,* 56, 690-700.

Goode, D. J. (1986). "Selection of soils for wick effect covers." *Symposium of. Geotech. & Geohydrol. Aspects of Waste Manage.,* Fort Collins, CO, 101- 109.

Hakonson, T. E., Maines, K. L., Warren, R. W., Bostick, K. V., Trujillo, G., Kent, J. S. and Lane, L. J. (1994). "Hydrologic Evaluation of Four Landfill Cover Designs at Hill Air Force Base, Utah", Los Alamos National Laboratory report LAUR-93-4469, prepared for US Department of Energy.

Nyhan, J. W., Hakonson, T. E. and Drennon, B. J. (1990). "A Water Balance Study of Two Landfill Cover Designs for Semiarid Regions." *J. Environ. Qual.,* 19(2), 281-288.

Ross, B. (1990). "The diversion capacity of capillary barriers." *Water Resour. Res.,* 26, 2625 - 2629.

Sherard, J. L., Dunnigan, L. P. and Talbot, J. R. (1984). "Basic Properties of Sand and Gravel Filters." *J. Geotech. Engng.,* 110, 684-700.

Steenhuis, T. S., Parlange, J.-Y. and Kung, K.-J. S. (1991). Comment on "The diversion capacity of capillary barriers" by B. Ross, *Water Resour. Res.,* 27, 2155-2156.

Stormont, J. C. (1995). "The effect of constant anisotropy on capillary barrier performance." *Water Resourc. Res.,* 31, 783-785.

van Genuchten, M. Th. (1980). "A Closed-form Equation for Predicting the Hydraulic Conductivity of Unsaturated Soils." *Soil Sci. Soc. Am. J.,* 44, 892-898.

Wing, N. R. (1993). "Permanent Isolation Surface Barrier: Functional Performance." Westinghouse Hanford Corporation, Hanford, WA, *Report WHC-EP-0650.*

Why Wet Landfills with Leachate Recirculation are Effective

By Debra R. Reinhart,[1] Member, ASCE

Abstract

Wet cell technology offers important advantages in the management and treatment of municipal solid waste, including accelerated waste stabilization rates, enhanced gas production, facilitated leachate management, volume reduction, and minimized long-term liability. These advantages have been documented in laboratory, pilot and full-scale investigations. Although challenges remain in implementing the technology, bioreactor landfills are designed and operated with increasing frequency.

Introduction

Regulatory and environmental expectations today demand a new approach to municipal solid waste (MSW) management, and in particular to landfill design and operation. The typical landfill of today is rapidly filled (as a result of high waste receipt rates and cellular design), tends to be quite deep, and is closed with an impermeable cap immediately after filling. These factors tend to limit moisture introduction, which is essential to the degradation of organic waste fractions. Without the benefit of adequate moisture, the modern landfill will serve primarily as a temporary storage device with only limited degradation. Once the environmental barriers (caps and liners) fail and permit moisture introduction, the consequential biological activity may result in gas and leachate production and, potentially, adverse environmental impact.

Under proper conditions, the rate of MSW biodegradation in a landfill can be stimulated, enhanced, and controlled within certain limits. Environmental conditions which most significantly impact biodegradation include pH, temperature, nutrients, absence of toxins, moisture content, particle size, and oxidation-reduction potential. Of these, the most critical parameters affecting MSW biodegradation has been found to be moisture content. Once the landfill has been capped, moisture content can be

[1]Associate Professor, Civil and Environmental Engineering Department, University of Central Florida, Orlando, Florida 32816-2450

most practically controlled via leachate recirculation. Leachate recirculation provides a means of optimizing environmental conditions within the landfill, providing enhanced stabilization of landfill contents as well as treatment of moisture moving through the fill. This paper will document the effectiveness of wet cell operation in five areas; treatment of wastes, leachate treatment and management, gas production enhancement, waste volume reduction, and long-term environmental impact.

Waste Treatment

The typical modern landfill is filled and closed within two to five years, and because of high construction costs, is built to depths of well over 30 to even 100 m, minimizing exposed surface area. In addition, many states prohibit the disposal of yard waste in landfills, eliminating a significant source of moisture. As a consequence, calculations show that even if emplaced waste captures every drop of precipitation in the wettest climates, moisture content of the waste at closure will be below optimum levels for biological degradation (Leszkiewicz and McAulay, 1995). In reality, much of the water entering a landfill finds highly permeable pathways to the collection system and is not absorbed by the waste. Once closed, further introduction of moisture is prevented by impermeable caps. As degradation proceeds, moisture content continues to decline with losses to biological uptake and vaporization, gas and waste degradation may slow further.

The addition of moisture has been demonstrated repeatedly to have a stimulating effect on waste degradation (Barlaz, et al, 1990), although some researchers indicate that it is the movement of moisture through the waste as much as water addition itself that is important (Klink and Ham, 1982). Full-scale documentation of enhanced waste degradation resulting from increased moisture content has been obtained by researchers at the Southwest landfill in Alachua County, Florida (Miller, et al, 1994). MSW samples were collected in areas experiencing leachate recirculation as well as dry control areas and examined for biological methane potential (BMP). A 50 percent decrease in BMP was measured in wet samples (46 percent wet basis) over a one year period. Negligible decreases in BMP were observed in dry samples (29 percent wet basis) over the same period, clearly demonstrating the enhancement of waste biodegradation as a result of leachate recirculation.

In wet climates, when large areas are open during early phases of landfill operation, large volumes of leachate will collect. Reintroduction of this moisture can help to maintain moisture content at optimum levels exceeding 50 percent (wet basis). In dryer climates, it may be necessary to introduce moisture from other sources such as municipal or industrial wastewater treatment facilities. Such practice currently is precluded by Subtitle D of RCRA and many state regulations, however, amendments of these regulations may be warranted to allow enhanced degradation and resulting benefits.

Proper design of leachate recirculation devices to provide for even and efficient distribution of moisture within the landfill is an essential (and often challenging) design criteria. In addition, efforts to open bags and break up the waste during placement and compaction can facilitate moisture movement by making the waste more homogeneous, as can the use of highly permeable daily and intermediate cover soils or alternate daily cover materials. Postponement of final RCRA closure should be consideration to allow for placement of an interim cap which provides for limited infiltration of moisture and maintenance of appropriate conditions for biodegradation of waste.

Leachate Management

Leachate management can be the most costly and difficult aspect of MSW landfilling. Landfills located in areas with wet climates and remote municipal wastewater treatment facilities are particularly problematic. In fact, when forced to transport leachate to treatment via truck, the danger of potential accidents can be the greatest liability that managers of active landfills face. Wet cell technology can provide a means of reducing leachate volume and improving leachate quality.

Analysis of leachate volumetric data from six full-scale recirculating landfill sites indicated that off-site disposal of leachate ranged from 0 to 59 percent of leachate generated (Reinhart, 1995). Data suggested that, unlike conventional landfills, the volume of leachate requiring off-site management at recirculating landfills was a function not only of the volume of leachate generated but was also related to the availability of on-site storage. At sites where large storage volumes were provided relative to the size of the landfill cell (greater than 700 m^3/ha), off-site management of leachate was minimized (frequently no off-site management was required for long periods of time). It was also observed that sites with relatively little storage were compelled to recirculate leachate at much higher rates than those with large storage volumes.

The relative efficacy of leachate recirculation in enhancing leachate treatment at full-scale is difficult to quantify, because of the lack of conventional/recirculation parallel operations. Recognizing this limitation, leachate Chemical Oxygen Demand (COD) data were gathered from the literature for conventional and recirculating landfills (Reinhart, 1995). Leachate organic strength usually increases early during the landfilling process due to hydrolysis and solubilization of organic material. Once methanogenesis is established, the leachate organic strength declines. The slope of the decline of COD was calculated by non-linear regression of chronological COD data. From this slope a COD half-life of approximately 10 years for conventional landfills was determined. A similar analysis of leachate COD for full-scale leachate recirculating landfills having moved to maturation phases was made. Here, half-life values of 230 to 380 days were observed. Clearly, recirculation significantly increased the rate of the disappearance of organic matter in leachate.

Leachate recirculation also has important consequences with regard to metal contamination of leachate. The primary removal mechanism for metals in conventionally operated landfills appears to be washout, although limited chemical precipitation may occur. In leachate recirculating landfills, the primary metal removal mechanisms appear to be sulfide and hydroxide precipitation. Gould, et al, (1989) found that leachate recirculation stimulated reducing conditions providing for the reduction of sulfate to sulfide, which moderated leachate metals to very low concentrations. Chian and DeWalle (1976) reported that the formation of metal sulfides under anaerobic conditions effectively eliminated the majority of heavy metals in leachate. In addition, under neutral or above neutral leachate conditions, promoted by leachate recirculation, metal hydroxide precipitation is enhanced. With time, moderate to high molecular weight humic-like substances are formed from waste organic matter in a process similar to soil humification. These substances tend to form strong complexes with heavy metals. In some instances, a remobilization of precipitated metals can result from such complexation once the organic content has been stabilized and oxic conditions begin to be reestablished (Pohland, et al, 1992). The potential for remobilization supports the idea of inactivating the landfill (removing all excess leachate) once the waste has been stabilized.

Where leachate treatment is the primary objective of wet cell technology, leachate recirculation may be confined to treatment zones located within the landfill where appropriate processes are optimized. Use of in-situ nitrification, denitrification, anaerobic fermentation, and methanogenesis have been proposed to treat leachate, depending on the phase and age of the waste (Pohland, 1995). For example, pilot studies in Sweden have successfully investigated a two step degradation process within a landfill, whereby acidogenic conditions were maintained in one portion of the landfill, and methanogenic conditions in another part. The high strength, low pH leachate produced within the acid phase area was recirculated to the methane producing area for treatment. Consequently, methane production was accomplished under controlled conditions in an area suitably designed for maximum methane recovery (Lagerkvist, 1991).

Gas Production Enhancement

Several laboratory and pilot-scale lysimeters have documented increased gas production rates and total yields as a result of moisture addition (Pohland, 1975; Pohland, et al, 1992; Buivid, et al, 1981). Limited data suggest that, as in lysimeters, gas production at larger sites is significantly enhanced as a result of both accelerated waste stabilization as well as the return of organic material in the leachate to the landfill for conversion to gas (as opposed to washout in conventional landfills). Parallel one-acre cells operated by the Delaware Solid Waste Authority comparing conventional operation and leachate recirculation showed a twelve-fold increase of gas production in the recirculating cell over that of the conventional cell (DSWA,

1993). Gas emission measurements made by researchers at a recirculating landfill in Alachua County, Florida, revealed a doubling of gas production rates from waste located in wet areas of the partially recirculating landfill relative to comparably aged waste in dry areas. This fact was corroborated by comparison of measurements of methane potential from samples obtained in wet and dry areas of the landfill (Miller, et al, 1994).

Gas production enhancement can have positive implications for energy production and environmental impact, however, only if gas is managed properly. The facility must be designed to anticipate stimulated gas production, providing efficient gas capture during active phases prior to final capping. Captured gas, in turn, must be utilized in a manner which controls the release of methane and nonmethane organic compounds and/or provides for beneficial offset of fossil fuel use.

Waste Volume Reduction

Another consequence of wet cell operation is enhanced subsidence rates. Studies investigating the impact of leachate recirculation on settling have shown that wet cell technology enhances the rate and extent of subsidence. At the Sonoma County, CA pilot-scale landfill, the leachate recirculated cell settled by as much as 20 percent of its waste depth, while dry cells settled less than 8 percent (Leckie, et al, 1979). Wet cells at the Mountain View Landfill, CA settled approximately 13 to 15 percent, while control dry cells settled only 8 to 12 percent over a four year period (Buivid, et al, 1981). Wetting of waste as it is placed has been practiced for many years as a method of increasing compaction efficiency (Magnuson, 1992). Rapid and predictable settlement can provide an opportunity to utilize valuable air space prior to closure of the cell. Enhanced degradation rates can provide a means to meet mandated waste volume reduction in some parts of the world. Landfill reclamation and final site use are also facilitated by timely volume reduction provided by moisture control. The difficulty and expense of maintaining the integrity of the final cap over the long-term can be reduced as well.

Long-Term Liability

Long-term liability concerns can be minimized if waste is quickly treated to a point where further degradation will not occur or will occur so slowly that leachate contamination and gas production are sufficiently negligible to no longer pose a threat to the environment. A design life of 20 yrs for geosynthetic membranes may not provide adequate protection for the conventional landfill with stabilization periods of many decades. The potential impact on groundwater from a cleaner leachate is significantly reduced. Similarly, gas production confined to a few years rather than decades provides opportunity for more efficient control and destruction of air toxics and greenhouse gases. With sufficient data, regulators may be convinced to reduce long-term gas, leachate, and groundwater monitoring frequency and duration for leachate recirculating landfills, recognizing the reduced potential for adverse environmental impact. Reduced liability (and associated costs required for financial assurance) and minimal monitoring would translate into significant cost savings.

Conclusions

Wet cell technology does make a difference relative to waste stabilization rates, gas production, leachate management, volume reduction, and long term liability; all critical issues for landfill designers and operators. Significant challenges remain for the full-scale implementation of this technology, including leachate distribution (integrity of trenches and wells), gas collection (particularly during active phases), regulatory issues (such as moisture allowances, sludge addition, leachate depth above liner, and closure mandates), and operator training (regarding operation of the landfill as a treatment process). With continued research efforts and the global use of wet cell technology, these challenges can be met, resulting in a new and far more safe and effective approach to landfill operation.

References

Barlaz, M.A., D. M. Schaefer, and R.K. Ham (1990) "Methane Production From Municipal Refuse: A Review of Enhancement Techniques and Microbial Dynamics," Critical Reviews in Environmental Control, 19 (6) 557-584.

Buivid, M.G., D.L. Wise, M.J. Blanchet, E.C. Remedios, B.M. Jenkins, W.F. Boyd, and J.G. Pacey (1981) "Fuel gas Enhancement by Controlled Landfilling of Municipal Solid Waste," Resources and Conservation, 6, 3.

Chian, E.S. and F.B. DeWalle (1976) "Sanitary Landfill Leachates and Their Treatment," J. Environmental Engineering Division ASCE, 102 (EE2) 411.

Delaware Solid Waste Authority (1993) "Test Cell Report, November, 1992 - May, 1993," Central Solid Waste Management Center, Sandtown, Delaware.

Gould, J. P., W.H. Cross, and F. G. Pohland (1989) "Factors Influencing Mobility of Toxic Metals in Landfills Operated with Leachate Recycle," proceedings of Emerging Technologies in Hazardous Waste Management, May 1-4, Atlanta, GA.

Klink, R.E. and R. K. Ham (1982) "Effect of Moisture Movement on Methane Production in Solid Waste Landfill Samples," Resources and Conservation, 8, 29.

Lagerkvist, A. (1991) "Two Step Degradation - An Alternative Management Technique," Proceedings Sardinia 91, Third International Landfill Symposium, Calgari, Italy, (October 14-18).

Leckie, J. O., J. G. Pacey, and C. Halvadakis (1979) "Landfill Management with Moisture Control," J. Environmental Engineering ASCE, 105 (EE2) 337.

Leszkiewicz, J. and P. McAulay (1995) "Municipal Solid Waste Landfill Bioreactor Technology Closure and Post-Closure," presented at the US EPA Seminar - Landfill Bioreactor Design and Operation, March 23-25, Wilmington, DE.

Magnuson, A., (1993) "Methane Gas Collection: Does it Pay," MSW Management, (March/April) 40.

Miller, W.L., T. Townsend, J. Earle, H. Lee, D.R. Reinhart (1994) "Leachate Recycle and the Augmentation of Biological Decomposition at Municipal Solid Waste Landfills," Presented at the Second Annual Research Symposium, Florida Center for Solid and Hazardous Waste Management, Tampa, Florida (October 12, 1994).

Pohland, F. G. (1995) "Landfill Bioreactors: Historical Perspective, Fundamental Principles, and New Horizons in Design and Operations," presented at the US EPA Seminar - Landfill Bioreactor Design and Operation, March 23-25, Wilmington, DE.

Pohland, F.G., W.H. Cross, J.P. Gould, D.R. Reinhart, (1992) "The Behavior and Assimilation of Organic Priority Pollutants Codisposed with Municipal Refuse," USEPA, EPA Coop. Agreement CR-812158, Volume 1.

Pohland, F. G. (1975) "Sanitary Landfill Stabilization with Leachate Recycle and Residual Treatment," EPA/600/2-75-043, US Environmental Protection Agency, Cincinnati, OH.

Reinhart, D.R. (1995) "Active Municipal Landfill Operation," draft report to US Environmental Protection Agency, Cincinnati, OH.

WHY DRY LANDFILLS WITHOUT LEACHATE
RECIRCULATION ARE EFFECTIVE

Robert E. Landreth[1]

Introduction

The municipal solid waste that is generated in the
U.S. continues to demand attention for its final
disposition. Data would suggest that approximately
two-thirds of this waste will end up in a sanitary
landfill. EPA has published and reviewed individual
state programs for acceptance into managing the solid
waste via landfilling. Most states are requiring the
landfill approach although thermal destruction,
recycling, and composting are gaining in popularity.

Current Landfill Designs

Currently, landfills must follow 40 CFR 257,
258[6]. These minimum standards include a generic liner
design. The generic liner design is a composite
liner - 60 cm compacted clay with a geomembrane and a
leachate collection system capable of maintaining less
than 30 cm of head. In approved States, States where
the landfill rules meet or exceed the Federal rules, an
alternative design may be used. The alternative design
includes a model where the maximum contaminant
concentration levels are below a specified level at the
point of compliance.

The cover design is based on the type of liner
system used. One such design is a geomembrane overlain
by a moisture collection system and soil capable of

---

[1]Chief, Municipal Solid Waste and Residuals Management
Branch, U.S. Environmental Protection Agency,
26 W. Martin Luther King Drive, Cincinnati, Ohio 45268

supporting vegetation. The geomembrane may be
underlain by natural soil or a geosynthetic clay liner
(GCL). This design should prevent the bathtub
experience that has been reported for landfills.
Provisions are made to collect the decomposition gases
vented to the atmosphere.

Discussion of Differences

    This section of the paper will be concerned with
the differences between the leachate recirculation
landfill concept and modern landfill disposal
practices. The differences are discussed in terms of
what is known to occur at modern landfills and what the
issue is with leachate recirculation.

    Timing will play a major role in determining
whether or not leachate recirculation can be performed.
In order to have recirculation, leachate collection
systems have to be designed to handle the increased
movement of liquids in the landfill. In recent studies
at Drexel University[1], the growth of organisms and
particulate clogging of geosynthetics has determined
that porosity of the geosynthetic is critical. The
Germans[2], have concluded that 16-32 mm cobbles are
necessary to keep the leachate collection system open
and flowing. Questions have been raised about how the
system should be designed and how long it should
operate. These decisions must be made while the
landfill is being designed. On the other hand a
leachate collection system, in a modern landfill, has
to remain in service for the life of a landfill, plus a
couple of years. Once the landfill is closed there
generally is only minimal infiltration depending on the
cover design, and thus only minimal amounts of
water/leachate to be collected.

    During the active life of the landfill, the HELP
(Hydrologic Evaluation of Landfill Performance) model[3]
can be used to conservatively estimate what amount of
leachate will be generated and the amount that needs to
be collected.

    When considering the leachate collection system,
the designer should review the designs for the
reinjection systems to ensure the two are compatible
and the sizes are correct. Will it be a pipe or pipe
with a nozzle, a wet well, spray irrigation, pond,
etc.? Research has not thoroughly investigated the
different types of reinjection systems to get the
leachate back into the landfill. Concerns that are

raised include the ability of the reinjection system to
remain free from clogging due to the large amount of
solids and chemicals in the leachate, the ability to
provide a uniform wetting front in the waste mass, the
physical spacing of the injection system, both
horizontally and vertically, and the amount of leachate
to add and what frequency. Field experience with these
systems will tell us of their durability as well as
their expected service life. These issues must be
solved if the leachate recirculation is to be
effective.

Research not only has to solve the above issues,
but the leachate injection system has to be evaluated
for long-term durability. The systems should be
designed for life times of 10-20 years. Landfill
leachate chemistry is very aggressive, has a changing
quality and quantity, and can reduce the effectiveness
of any injection system. The leachate chemistry effect
on long-term operation must be assessed.

The leachate reinjection system will play an
important role in determining whether the entire waste
mass will become wetted and subject to deterioration.
Short circuiting will be common which will prevent the
waste mass from becoming wet. If the operator applies
more leachate than what the landfill can properly
utilize, the system may become "sour" or upset. How
long does one have to wait and under what conditions
does one restart the physical, chemical, and biological
decomposition?

The waste mass to be considered is the entire
landfill. Without treating the whole mass, complete
stabilization of the facility will not occur. One of
the biggest questions remains as to how to wet the
entire facility or at least a major portion of waste
mass. Part of the problem is caused by the plastic
bags used to contain individual household waste. It is
assumed that these bags are split when compacted either
in the collection truck or in the landfill. In reality
most of the bags are split, but to ensure that they are
entirely split, a spotter at the working face may have
to be used. The spotter becomes at risk at very active
landfills and splitting of bags can become a very
dangerous job. Splitting of bags is a non-issue at
modern landfills and may in fact add to the overall
landfill containment system. These unsplit bags may
house liquids or semi-solids that are contained and are
not available for mixing with the rest of the waste
mass.

The daily cover used at the site will generally
isolate the waste for that day. The soils generally
used have important functions. These include reduction
of fire potential, the reduction of blowing debris,
flies, birds, and smell. Both the modern landfill and
the recirculation landfill would have to provide for
this service. Geosynthetics have been showing promise
for daily cover[4], [5]. Included in this category are
chemical foams, geosynthetic sheets, and other
materials. The geosynthetic materials generally take
less air space, can be used more than once, and do not
inhibit the movement of moisture (good for both
landfill types). Alternative daily cover materials are
somewhat difficult to place in inclement weather and
not all States accept these materials as a substitute
for the soils. Whatever the daily cover is it will
most likely decrease the effectiveness of leachate
recirculation. Landfills that have been exhumed have
had a substantial amount of soil. Soil has also been
shown to cause micro-barriers within the landfill.
Alternative material may also provide a barrier
although this would be of less concern than soil.

Landfill wastes are generally compacted to about
5.8 to 9.3 kN/m^3, depending on waste moisture, type of
equipment, etc. In modern landfills, the objective is
to maximize compaction. The reasons are simple:
maximize the compaction to increase free air space and
to minimize moisture movement thus reduce settlement
and subsidence. In a recirculation type landfill
research on injection nozzles and the associated
plumbing need to be optimized as well as the compaction
of the waste. If waste is compacted as in a modern
landfill, the leachate might not trickle through the
waste mass. If compacted lightly, significant amounts
of settlement/subsidence can be expected. Also with
settlement/subsidence comes the strong possibility of
breaking the leachate injection lines. One possibility
to minimize the impact of settlement/subsidence is to
install vertical injection wells. However, these
systems would have to be optimized for injection
frequency, spacing, size, etc.

The landfill operator is considered to be a key
person for both the modern or recirculation type of
landfill. His presence can make the difference between
a smooth operation and one that is haphazardly
operated. In addition to the modern landfill duties,
the operator should be knowledgeable in hydraulics for
the leachate injection system; knowledgeable in
landfill operation for daily cover, compaction of the
waste, when and how much, and where within the landfill

to inject leachate to optimize the process;
knowledgeable in leachate chemistry to understand when
the leachate is about to become toxic and how, when and
what to remove from the leachate; knowledgeable in the
leachate nutrients; knowledgeable in gas generation and
how to effectively remove the gas; and a states-person
to discuss all of these issues with interested people
who will visit the site to review its operation.

     Leachate recirculation is not without its list of
items for research. Significant resources will be
required just to answer the few questions raised here.
Research is happening and will eventually get to the
solutions. Research may also answer the long-term
question of "When is the leachate recirculation process
completed?" What are the options for the site when
recirculation is completed? One strong possibility is
to mine the site for recyclables and then reuse the
site again. This may eliminate the need for acquiring
a permit and would allow for the installation of modern
containment technology. Another option is to walk away
from the site since it should be physically,
chemically, and biologically stabilized.

     On the other hand current landfilling techniques
look reasonably straight forward with solutions to most
problems. Why should it not work? Landfilling has
been around for some time and most of the answers are
known. Modern designs are successfully being
implemented throughout the country.

Summary

     Modern landfilling practices can be successfully
implemented throughout the country. The Environmental
Protection Agency has designs that are being
implemented and landfills are being constructed,
operated, and closed. Problems that may be encountered
are easily understood and corrected.

     Leachate recirculation is a concept to accelerate
the physical, chemical, and biological stabilization of
a landfill facility. A number of issues still remain
including the collection and reinjection systems, how
to maximize the wetting of the landfill mass, the
compaction of the waste, and the chemistry of the
landfill leachate.

     The major problem for researchers is that leachate
recirculation needs to be field evaluated under
different hydrogeologic environments. This in turn

costs significantly and is time consuming. Changes in
design may take additional time to correct. On the
other hand, the modern landfill has proven technology
and creates minimal problems.

References

1. U.S. EPA 1995. "Leachate Clogging Assessment of
    Geotextile (and soil) Landfill Filters,"
    Koerner, R.M. and G.R. Koerner, Drexel University,
    CR819371, In Press, Risk Reduction Engineering
    Laboratory, Cincinnati, Ohio 45268.

2. "Geosynthetic Liner Systems: Innovations,
    Concerns, and Designs," Koerner, R.M. Proceedings
    of the 7th GRI Seminar, Decemberr 14-15, 1993,
    Philadelphia, PA.

3. U.S. EPA 1995. "Hydrologic Evaluation of Landfill
    Performance (HELP) Model," Version 3.04,
    Schroeder, P., U.S. Army Engineers - Waterways
    Experiment Station. IAG-DW96936492 (EPA/600/R-
    94/168a/b), Risk Reduction Engineering Laboratory,
    Cincinnati, Ohio 45268, September 1994.

4. U.S. EPA 1992. "Alternative Daily Cover Materials
    for Municipal Solid Waste Landfills," PRC
    Environmental Management, Inc., CR68-W9-0041,
    WA#R2919, Region XI, Office of Solid Waste, U.S.
    Environmental Protection Agency, 1992.

5. U.S. EPA 1993. "The Use of Alternative Materials
    for Daily Cover at Municipal Solid Waste
    Landfills," Pohland, F.G. and J.T. Graven,
    University of Pittsburgh, EPA/600/R-93/172, NTIS
    PB 93-227197, Risk Reduciton Engineering
    Laboratory, Cincinnati, Ohio 45268, July 1993.

6. 40 CFR 257 and 258, Federal Register, Part II,
    Environmental Protection Agency 40 CFR Parts 257
    and 258, Solid Waste Disposal Facility Criteria -
    Final Rule, pages 50978-51119, Vol 56, No. 196,
    October 9, 1991.

Comparison and Contrast of Leachate Collection Systems
for
Subtitle D and Non-Subtitle D Landfills
by
A.S. Dellinger; Member, ASCE[1] and Roger Strohm [2]

ABSTRACT

Management of leachate generated by landfills is necessary to prevent environmental degradation. However, the techniques for managing leachate vary from landfill to landfill and what may be successful at one site cannot necessarily be incorporated at another.

This paper details the structures and design parameters used in managing and controlling leachate at landfill sites that are designed for Subtitle D compliance as well as strategies used to manage leachate at sites without liner systems. A comparison and contrast between two landfills managed by the City of Los Angeles will be presented, including discussion of each site's leachate collection and removal system strengths and limitations.

LEACHATE

Leachate is defined as a strong wastewater that is generated when precipitation from rainwater infiltration, overland flows across the site, or sub-surface groundwater flows, saturates the refuse mass. Refuse emplaced within landfills contains moisture. Leachate is produced when water from either infiltration or through sub-surface groundwater flows comes into contact with this waste and exceeds its field capacity. It should be noted that many people believe there is not any difference between field capacity and the saturation point, and that both remain constant under static conditions - such is not the case. The internal mechanics of the landfill are dynamic processes and as refuse decomposes and consolidates, its leachate retention level decreases. Thus leachate can be generated in otherwise dry or arid environments where the groundwater table is located far below the landfill.

---

[1]Professional Engineer, City of Los Angeles, Bureau of Sanitation, 419 S. Spring Street, Suite 800, Los Angeles, CA 90013

[2]Professional Engineer, City of Los Angeles, Lopez Canyon Landfill, 11950 Lopez Canyon Road, Lake View Terrace, CA 91342

SUBTITLE D

In response to concerns over off-site migration of leachate and its contamination of the underlying groundwater aquifer, the United States Environmental Protection Agency (EPA) issued 40 CFR, Part 258, commonly referred to as Subtitle D on October 9, 1991. The regulations required that landfills meet certain design standards with the specific intent of encapsulating refuse. Minimum design standards for landfills complying with this statute include a composite liner system consisting of clay with hydraulic conductivity less than 1×10^{-7} cm/sec and geosynthetic layers. A drainage layer is also required within the liner system to collect leachate within the landfill before escaping into the underlying aquifer.

In conjunction with Subtitle D, the quantity of leachate produced on-site can be restricted by limiting the acceptance of saturated refuse waste streams and the construction of landfill cover systems which limit infiltration from overland water flows and precipitation. An adequately designed cover system also serves as a barrier to leachate surface seeps from the landfill and prevents exposure to the environment.

TOYON CANYON: A NON-SUBTITLE D LANDFILL

Landfills constructed prior to passage of Subtitle D regulations present special challenges in addressing the leachate issue and remediating its hazards. The Toyon Canyon Sanitary Landfill is operated by the City of Los Angeles within a public park. Approximately 14.5 million metric tons (16 million tons) of refuse were disposed at this site from 1957 until 1985 in an area covering 0.36 square kilometers (90 acres). The site is a canyon fill with average refuse depths of 88 m (290 feet). Natural springs are located throughout the park and actually inundate the landfill.

The liner system and subdrains at this site are non-existent. Retrofitting the landfill to satisfy Subtitle D requirements is cost prohibitive and not required under Subtitle D. However, the Bureau did implement several systems designed to mitigate off-site migration of leachate through groundwater infiltration and surface seepage as detailed below.

Initial Installation

In 1979, leachate was suspected of migrating off-site at the toe of the landfill with the potential danger of impacting the park . To remedy this hazard, a 0.6 m (2 feet) wide soil-cement barrier wall and a 0.9 m (3 feet) wide french drain (Figure 1) were constructed in 1981. The barrier wall is keyed 15 cm (6 inches) into bedrock and lined with a 0.75 mm (30 mil) polyethylene liner. On the side facing the landfill, the french drain is lined with a permeable geofabric that allows leachate to enter the trench while restricting the inflow of sediment. The top 1.4 m (4.5 feet) of the trench is covered with natural soil; the bottom of the trench, ranging from 0.2 to 3.5 m (0.5 to 11.5 feet) in depth, consists of 4 cm (1-1/2 inch) gravel surrounding a 15 cm (6 inch) polyvinyl chloride (PVC) pipe. This pipe prevents lateral transmission of leachate, routing it, instead, to either of two manholes constructed at each trench end. From these manholes, the leachate is conveyed to the sanitary sewer system.

In 1981, leachate began seeping from the lower six benches. To prevent exposure and surface run-off of leachate (with potential contamination of stormwater run-off), a leachate collection and removal system (LCRS) was designed. This system consisted of 10 cm (4 inch) diameter PVC pipes which collected leachate along each bench to a central manhole and through a 10 cm (4 inch) PVC pipe from bench-to-bench, before tying into the sanitary sewer system for disposal and treatment.

For each of the lateral runs along the benches, the perforated PVC pipe was placed in a trench approximately 0.9 m (3 feet) wide by 1.5 m (5 feet) deep and surrounded with gravel to a height of 0.9 m (3 feet); the remaining top 0.6 m (2 feet) were covered with clean fill material (typically clay). As with the barrier cut-off wall, the trench was lined with a permeable geofabric to reduce the amount of sediment inflow. The pipes were perforated by drilling 2 cm (3/4 inch) holes, spaced at 15 cm (6 inch) intervals along the pipe length.

To collect and contain leachate seeps in the south-western corner of the landfill, a 24 m (80 feet) long trench, 0.6 m (2 feet) wide by 1.8 m (6 feet) deep, was installed. As with the front face LCRS, a 10 cm (4 inch) PVC perforated pipe was placed and backfilled with 1.2 m (4 feet) of 3.8 cm (1-1/2 inch) gravel and covered with 0.6 m (2 feet) of natural soil. The sides were lined with a 0.25 mm (10 mil) polyethylene liner that channeled leachate to a 3785 liter (1000 gallon) holding tank buried 1.8 m (6 feet) underground.

LCRS Upgrade
Leachate continued to seep from the slopes above benches five and six. To remediate this problem, gravel subdrains were installed from the point of discharge, down to the toe of slope and connecting with the existing LCRS gravel subdrain. These additions eliminated leachate seeps as they occurred, minimizing erosion and odor problems.

When they occur, additional leachate seeps are covered with clay and compacted. If the seep persists, then the soil where the seep occurs is excavated and disposed at an active landfill. The excavated area is then filled with a 2.5 cm (1 inch) washed aggregate before covering with clay.

To further facilitate leachate collection and alleviate the pressure gradient, leachate is pumped from several landfill gas (LFG) collection wells and routed to a LCRS bench manhole.

Leachate Generation Rates
In 1987, leachate generation from the front face system ranged from 11 to 27 liters per minute (3 to 7 gpm); November 1993 measurements found that leachate flowed at 23 liters per minute (6.1 gpm).

Maintenance Activities and Procedures
The installation of the LCRS initially solved the problem of leachate seeps from the landfill's face. However, after a period of four years, seeps were again observed. Additionally, down-drains conveying leachate from bench to bench occasionally experienced overflow at the manhole. After these events, the Bureau initiated a LCRS Maintenance Program.

The principal common form of leachate blockage resulted from carbonaceous growth caused by precipitation from the mild alkalinity associated with leachate. Over time, the carbonaceous substance builds up along the inside surface of the PVC pipe, constricting and eventually blocking flow; leachate backs up into the upper manhole and overflows. To effectively remove this growth, pipes require cleaning from bench-to-bench on a quarterly basis. A rodding machine with a root cutting head is inserted into the 10 cm (4 inch) PVC lines and channels through blockage from one bench to the other.

The leachate environment is physically harsh on PVC pipes and degrades its integrity. Further, the landfill's internal temperature is greater than 43.3° C (110°F). Refuse decomposition within the landfill causes settlement, which in turn, induces stress on the LCRS. The combination of these factors eventually results in the cracking and breakage of the pipe. As leachate may continue flowing through the surrounding gravel pack to a down-drain manhole or is reintroduced into the landfill, these breaks can go undetected until routine maintenance reveals their presence. The break is located by recording how much cleaning rod is inserted into the manhole before encountering resistance. At this location, maintenance crews must excavate down to the PVC pipe, cut and repair the pipe, before backfilling and recompaction. This location method is field-proven to minimize excavation time and worker exposure to buried refuse. As with seepage repairs, excavated material is hauled off-site for disposal at an active landfill.

Due to limitations in original design construction, only the down-drains from bench-to-bench can be repaired. The laterals installed along the slopes cannot be maintained as an adequate number of clean-outs were not installed.

A complementary maintenance procedure to roto-rootering is hydroflushing which involves inserting a hose into the leachate down-drain, and forcibly jetting water down the drain pipe to the next bench. The water's speed is greater than the scouring velocity and any particulates/fines which precipitate are flushed to the next bench where a vacuum/suction truck pumps them out.

Future

A geocomposite liner will be installed as part of the final closure design for this landfill. The liner, along with two soil barriers, each 0.3 m (1 foot) thick, will effectively eliminate surface water infiltration into the landfill. Consequently, any leachate that is generated will be from underlying groundwater flows inundating the refuse.

A detailed slope stability analysis was required due to the steep side slopes (2 horizontal to 1 vertical) and the static and dynamic loading from earthquakes. This analysis revealed that the factor of safety decreases where the foundation soils approach saturation. That is, a fully saturated landfill surface is not as stable, safe, or secure as one which is dry. Consequently, the Bureau has a significant need to ensure that leachate does not seep from the surface after final closure.

To prevent the leachate from infiltrating through the final cover, a subsurface drainage system will be installed (Figure 2). This system will consist of a geocomposite drainage layer on the front face, covering the slope surface and providing a conduit to the bench for any leachate. PVC or high-density polyethylene (HDPE) pipes, 20 cm (8 inch) in diameter, will channel the leachate along each bench to a down-drain where it will be routed in 20 cm (8 inch) pipes from bench to bench before eventual treatment and discharge.

To aid in leachate collection, the pipes along each bench will be placed in a 61 x 61 cm (24 x 24 inch) gravel filled trench which will be lined with a 226 g (8-ounce) geotextile trench liner and a 0.75 mm (30 mil) PVC trench liner (Figure 3). The geotextile trench liner will restrict the intrusion of fines into the collector pipes, thereby reducing maintenance needs. The PVC trench liner ensures collection and containment of leachate within the perforated pipe. To aid in maintenance of this system, clean-out ports will be installed every 61 m (200 feet). These ports also provide for simultaneous landfill gas extraction from the leachate collection system.

Current designs for the leachate calls for its disposal to the sanitary sewer system under existing discharge permits. However, feasibility studies are being explored for treating the leachate through filtration, reverse osmosis, flocculation, and/or clarification.

SUBTITLE D LANDFILL - LOPEZ CANYON SANITARY LANDFILL

The Lopez Canyon Landfill is an operating canyon landfill covering approximately 0.7 square km (170 acres); refuse disposal operations began in 1975. The current disposal area covers 0.21 square km (53 acres) and is constructed to Subtitle D standards. The canyon walls are typically 1:1 horizontal to vertical slopes with benches 4.6 m (15 feet) in width every 12.2 vertical m (40 vertical feet). Approximately 3630 metric tons (4000 tons) per day are disposed at the landfill.

The Subtitle D liner system at Lopez Canyon Landfill consists of a subdrain, a composite liner, a leachate collection system and a geotextile filter. These components work together as a system to prevent leachate from contaminating the groundwater.

Subdrain
The subdrain system on the bottom of the landfill has a different function and design than that on the slopes. On the floor, the primary purpose of the subdrain system is to act as a leachate leakage indicator. Its secondary purpose is to prevent hydrostatic pressure from building up beneath the liner. The bottom of the landfill is sufficiently higher than the top of the groundwater table so capillary rise is not an issue. However, excavation of the floor encountered several seeps from perched groundwater. The subdrain was routed through these seeps during construction to prevent hydrostatic pressure from building beneath the liner.

The floor subdrain design consists of a 15 cm (6 inch) perforated pipe placed in a 1.2 m deep by 0.6 m wide (46 x 24 inches) trench. The trench is backfilled with gravel which has a minimum hydraulic conductivity of 1x10⁻² cm/sec; a geotextile filter is wrapped around the gravel.

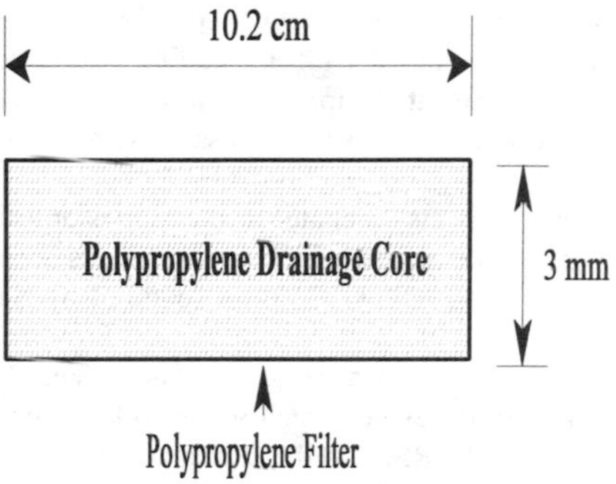

FIGURE 4 -Typical Wick Drain Cross-Section

The purpose of the subdrain on the slopes is to prevent hydrostatic pressure from building up beneath the liner. The slope subdrain design consists of wick drains[3] (Figure 4) overlain by a 340 g (12 ounce) filter geotextile. These wick drains are laid on the slope surfaces except where crossing the benches. Here, they are placed into trenches. These drains eventually tie into the floor subdrain system.

Liquid from the subdrain system is sampled semi-annually and compared against leachate analysis to determine if leachate is leaking through the liner and potentially contaminating the groundwater aquifer.

---

[3]Wick drains are defined as a two-part prefabricated geocomposite vertical soil drain consisting of a formed polypropylene core covered with a non-woven polypropylene filter fabric allowing water to pass into the drain core while restricting the infiltration of soil particles which clog the cores.

Composite Liner
The composite liner system is primarily responsible for containing leachate and preventing its migration off-site. The composite liner on the floor consists of a 0.6m (2 feet) thick earth liner with a hydraulic conductivity of $1x10^{-7}$ cm/s and overlain by a textured 2 mm (80 mil) HDPE flexible membrane.

On the slopes, a geosynthetic clay liner with a specified hydraulic conductivity of $1x10^{-9}$ cm/s is overlain by a textured 2 mm (80 mil) HDPE flexible membrane.

These composite liners limit the flow of leachate through the liner allowing the leachate to be collected in the leachate collection and removal system.

Leachate Collection and Removal System
To remove the hydrostatic pressure on the liner and prevent the hydraulic head from exceeding 30 cm, a leachate collection and removal system is installed above the floor and slope liners (Figure 5). The floor LCRS consists of a 30 cm (1 foot) thick gravel blanket with a specified hydraulic conductivity of $1x10^{-2}$ cm/s and a dendritic perforated pipe system which collects and routes the leachate from the gravel to the storage and disposal system.

A geonet collects leachate on the slopes and ties into the floor LCRS (Figure 6). The geonet, combined with the steep slopes at the Lopez Canyon Landfill, keeps the hydraulic head below 30 cm as required by Subtitle D.

Above the LCRS is a geotextile filter. This filter on the floor consists of an 227 g (8 ounce) geotextile, which serves to keep the 0.6 m (2 feet) operation layer from clogging the LCRS. On the slopes, a 340 g (12 ounce) geotextile is used to prevent the operation layer from clogging the geonet.

Leachate Disposal
Currently, leachate is routed from the LCRS to a sampling manhole outside the refuse boundary. The subdrain system runs to a separate sampling manhole at the same location. Based on quarterly sampling results, the leachate is either sent to the local sewer system or to storage tanks for treatment. Currently, all leachate is disposed into the sewer system.

Table 1 provides a constituent comparison between the leachate, subdrain liquid, and stormwater runoff for the Lopez Canyon Landfill as well as leachate analysis for the Toyon Canyon Landfill. Currently there is no indication of leaks. The stormwater runoff analyses is compared to the leachate and subdrain analysis as stormwater infiltrates the LCRS through areas of the geonet not yet covered with refuse. Leachate generation rates were measured as 1 liter per minute (0.25 gpm) on March 18, 1994. Following 3 cm (1.1 inches) of rain on March 25, 1994, the leachate flow rate was measured at 113 liters per minute (30 gpm), illustrating leachate generation from rainfall infiltration into the geonet.

TABLE 1 - Comparison of Leachate, Subdrain and Storm Water for Nonsubtitle D and Subtitle D Landfills

COMPOUND (ppb)	Toyon Landfill	Lopez Canyon Landfill		
	10/6/94 Leachate	11/10/94 Storm Water	12/07/94 Leachate	8/02/94 Subdrain
Methylene Chloride	2.96	2.61	110	<2.46
CIS-1,2-dichloroethene	<0.5	<0.60	2.24	<0.60
Chloroform	5.85	<0.23	0.24	<0.23
Benzene	3.42	<0.18	4.03	<0.18
Toluene	1.38	<0.30	8.06	<0.30
Tetrachloroethene	<0.5	<0.35	0.64	<0.35
Xylene	0.92	<0.22	0.97	<0.22
1,2,3 Trichlorobenzene	<0.5	<0.71	1.01	<0.71
Phenol	<1.0	<1.00	100	<1.00
bis-phthalate	96.9	<1.00	4.40	43.6
pH	7.6	7.3	4.11	7.0
Chloride mg/l	3470	48	69	75
Magnesium	150	62	127	not tested

Leachate Reduction

The City of Los Angeles takes several steps to reduce the quantity of leachate and improve its quality at the Lopez Canyon Landfill. The City stopped accepting sludge, an extremely wet waste, from its wastewater treatment plants, thereby reducing the volume of leachate produced while improving its quality. The City's collection crews have been trained to recognize household hazardous waste and set those items aside when picking up refuse, reducing the amount of hazardous waste disposed at the landfill. The City randomly inspects five loads of refuse per day at the landfill for hazardous waste. The worst routes are inspected again on their next trip. If a route is consistently high in the amount of collected hazardous waste, flyers and other educational materials are mailed out to households on that route to curtail hazardous waste disposal. The City also operates a mobile household hazardous waste collection vehicle to facilitate hazardous waste collection before disposal into the landfill.

To reduce rainfall infiltration, the City covers the trash daily with 30 cm (1 foot) of soil, 15 cm (6 inches) more than that required under Subtitle D. By minimizing the lengths of flowlines over the landfill decks and maintaining slopes of at least 3%, stormwater-infiltration is reduced. Also, no run-off outside of the refuse boundary is allowed to run onto and over the landfill. As part of the final cover design, a flexible membrane liner will be placed on the top deck with a 0.6 m (2 feet) thick clay liner on the 2:1 slopes further restricting water infiltration.

Maintenance
Simultaneous disposal of refuse during liner construction presented several challenges. For example, to simplify construction scheduling around disposal operations, the entire area (0.8 square km, 20 acres) to be filled during the landfill's current permit was lined. However, this left a large section, approximately 32000 sq. m (8 acres) of the slope liner exposed to rain. Rainfall on this portion of the liner was collected by the geonet and disposed.

The geonet also provides a pathway for landfill gas migration. This was resolved by locating horizontal gas collection wells 3.0 to 4.6 m (10 to 15 feet) away from the slope-trash interface, thereby eliminating surface emissions.

Title 23 of the California Code of Regulations states in part, "Leachate collection and removal systems shall be designed and operated to function without clogging... The system shall be tested at least annually to demonstrate proper operation." This was accomplished at the Lopez Canyon Landfill by installing cleanout pipes, which are connected to the LCRS piping system. However, these pipes cause disposal problems because they daylight into the trash disposal area and have been knocked over, buried, and crushed. A construction crew has been assigned to maintain and recover these pipes, resulting in increased operational costs.

CONCLUSION
All landfills generate leachate. With today's regulations, leachate is prevented from contaminating the groundwater aquifer by essentially entrapping the leachate in the landfill before subsequent collection and removal through drainage pipes. Landfills operating prior to the advent of liner systems pose a unique problem which can be resolved through a variety of creative means. For example, landfill leachate can be actively removed by installing dewatering wells; this solution is costly and often difficult as leachate may be perched within the landfill. Installing and maintaining a LCRS is also a viable solution for many landfill sites. If the groundwater table infiltrates the landfill, as with the Toyon Landfill, then collection of leachate through a geocomposite drainage layer as it seeps to the surface or at each individual toe slope is a practical and effective approach. Economic factors must also be explored. A landfill which will not undergo closure for some time may require installation of the LCRS as opposed to a geocomposite drainage layer combination, realizing significant financial savings. Operating landfills are now designed to comply with Subtitle D requirements with the intent of minimizing future leachate hazards. Operating municipal waste landfills should also take steps during refuse collection to reduce potential leachate hazards. Each of these actions serve to minimize the quantity of generated leachate.

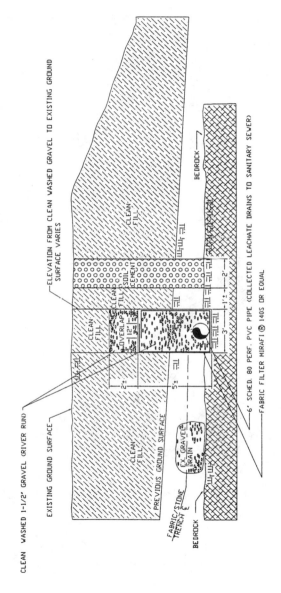

TOYON CANYON SANITARY LANDFILL
BARRIER CUT-OFF WALL CROSS-SECTION

FIGURE 1

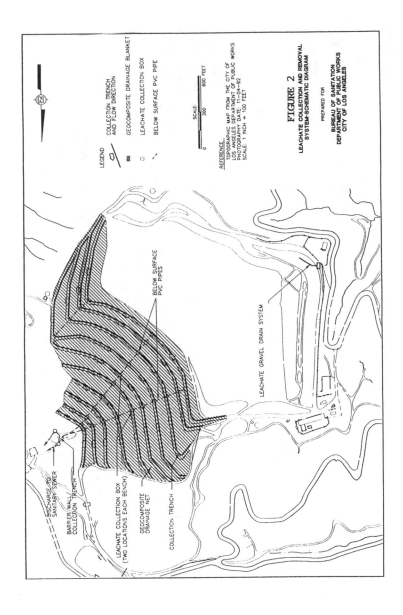

LEGEND

COLLECTION TRENCH AND FLOW DIRECTION

GEOCOMPOSITE DRAINAGE BLANKET

LEACHATE COLLECTION BOX

BELOW SURFACE PVC PIPE

SCALE
0   300   600 FEET

REFERENCE:
TOPOGRAPHIC MAP FROM THE CITY OF
LOS ANGELES DEPARTMENT OF PUBLIC WORKS
PHOTOGRAPHY DATE: 11-04-92
SCALE: 1 INCH = 100 FEET

FIGURE 2
LEACHATE COLLECTION AND REMOVAL
SYSTEM-SCHEMATIC DIAGRAM

PREPARED FOR

BUREAU OF SANITATION
DEPARTMENT OF PUBLIC WORKS
CITY OF LOS ANGELES

BELOW SURFACE
PVC PIPES

LEACHATE GRAVEL DRAIN SYSTEM

DISCHARGE TO
SANITARY SEWER

BARRIER WALL/
COLLECTION TRENCH

LEACHATE COLLECTION BOX
(TWO LOCATIONS EACH BENCH)

GEOCOMPOSITE
DRAINAGE NET

COLLECTION TRENCH

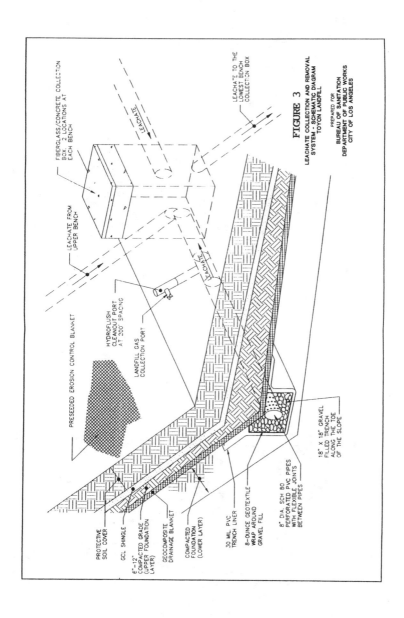

FIGURE 3

LEACHATE COLLECTION AND REMOVAL
SYSTEM - SCHEMATIC DIAGRAM
TOYON LANDFILL

PREPARED FOR
BUREAU OF SANITATION
DEPARTMENT OF PUBLIC WORKS
CITY OF LOS ANGELES

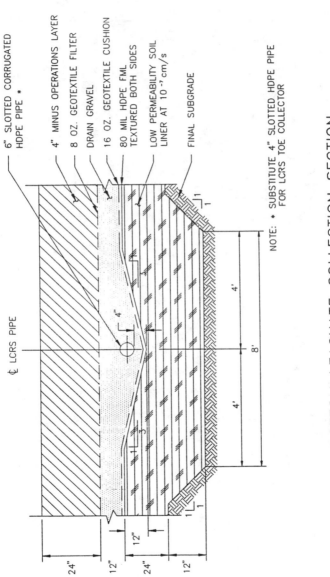

TYPICAL BOTTOM LEACHATE COLLECTION SECTION
LOPEZ CANYON LANDFILL

N.T.S.

FIGURE 4

NOTE: * SUBSTITUTE 4" SLOTTED HDPE PIPE
FOR LCRS TOE COLLECTOR

6" SLOTTED CORRUGATED HDPE PIPE *

4" MINUS OPERATIONS LAYER

8 OZ. GEOTEXTILE FILTER

DRAIN GRAVEL

16 OZ. GEOTEXTILE CUSHION

80 MIL HDPE FML TEXTURED BOTH SIDES

LOW PERMEABILITY SOIL LINER AT 10^{-7} cm/s

FINAL SUBGRADE

℄ LCRS PIPE

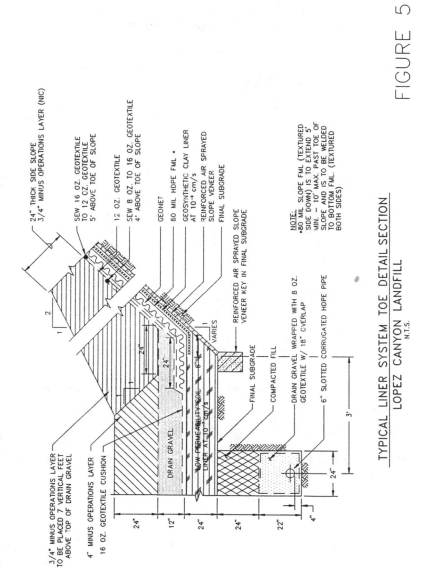

FIGURE 5

TYPICAL LINER SYSTEM TOE DETAIL SECTION
LOPEZ CANYON LANDFILL
N.T.S.

24" THICK SIDE SLOPE
3/4" MINUS OPERATIONS LAYER (NIC)

SEW 16 OZ. GEOTEXTILE
TO 12 OZ. GEOTEXTILE
5' ABOVE TOE OF SLOPE

12 OZ. GEOTEXTILE

SEW 8 OZ. TO 16 OZ. GEOTEXTILE
4' ABOVE TOE OF SLOPE

GEONET

80 MIL HDPE FML *

GEOSYNTHETIC CLAY LINER
AT 10⁻⁸ cm/s

REINFORCED AIR SPRAYED
SLOPE VENEER

FINAL SUBGRADE

REINFORCED AIR SPRAYED SLOPE
VENEER KEY IN FINAL SUBGRADE

FINAL SUBGRADE

COMPACTED FILL

DRAIN GRAVEL WRAPPED WITH 8 OZ.
GEOTEXTILE W/ 18" OVERLAP

6" SLOTTED CORRUGATED HDPE PIPE

3/4" MINUS OPERATIONS LAYER
TO BE PLACED 7 VERTICAL FEET
ABOVE TOP OF DRAIN GRAVEL

4" MINUS OPERATIONS LAYER

16 OZ. GEOTEXTILE CUSHION

DRAIN GRAVEL

LOW PERMEABILITY SOIL
LINER AT 10⁻⁶ cm/s

VARIES

24"

24"

6"

24"

12"

24"

24"

22"

4"

24"

3'

NOTE:
*80 MIL SLOPE FML (TEXTURED
SIDE DOWN) IS TO EXTEND 5'
MIN. - 10' MAX. PAST TOE OF
SLOPE AND IS TO BE WELDED
TO BOTTOM FML. (TEXTURED
BOTH SIDES)

References

American Wick Drain Corporation, Amerdrain 407 Vertical Soil Drain Product Literature, AMER 407-0293.

Barclays California Code of Regulations, Title 23, Sub-chapter 15.

Dellinger, A.S., Jeffrey G. Dobrowolski, and Constantin Pano, "In-situ Leachate Collection and Removal System." Seventeenth International Madison Waste Conference, September, 1994.

I.T. Corporation, 1988, "Solid Waste Assessment Test (SWAT), Toyon Landfill," prepared for the Bureau of Sanitation, Department of Public Works, City of Los Angeles, June.

I.T. Corporation, 1993, "Draft Report: Final Closure and Postclosure Maintenance Plan for Toyon Canyon Sanitary Landfill," prepared for the Bureau of Sanitation, Department of Public Works, City of Los Angeles, February.

I.T. Corporation, 1993, "Response to Regulatory Comments on Draft Report: Final Closure and Postclosure Maintenance Plan for Toyon Canyon Sanitary Landfill," prepared for the Bureau of Sanitation, Department of Public Works, City of Los Angeles, October.

Los Angeles Department of Public Works, Bureau of Sanitation, 1994, "Toyon Canyon Landfill Monitoring and Reporting Program for the Fourth Quarter, 1994," submitted to the California Regional Water Quality Control Board, December 15.

Los Angeles Department of Public Works, Bureau of Sanitation, "Lopez Canyon Landfill Monthly Reports," 1992-94.

Los Angeles Department of Public Works, Bureau of Sanitation, "Toyon Canyon Landfill Monthly Reports," 1992-94.

Report of Design for Lopez Canyon Landfill, Law Environmental, 1991.

Specifications for Liner Construction, Disposal Area "C" Phase II, City of Los Angeles, 1993.

Identifying Limitations on Use of the HELP Model

William E. Fleenor,[1] Student Member, ASCE and
Ian P. King,[2] Associate Member, ASCE

Abstract

Designs involved with operation and closure of landfills require an understanding of the movement of water through the waste matrix to predict the amount of moisture transported and to prepare for its collection and treatment. The Environmental Protection Agency (EPA) sponsored model, Hydrologic Evaluation of Landfill Performance (HELP), is routinely utilized in these efforts.

This report examines the ability of HELP to predict vertical moisture transport in order to identify any limitations of the model. The empirical algorithm of vertical moisture transport used in the HELP model was isolated and examined under three climate conditions varying from arid to humid. A finite element solution of the Richards equation, the physical equation describing unsaturated flow, was used to analyze the simplified vertical transport algorithm in HELP.

When compared to a Richards equation solution, HELP simulated barrier layer flux well under humid climate conditions, but the results demonstrated a trend of over-estimation in the model's predictions for arid and semi-arid conditions. As climate conditions became more arid, consistently greater over-predictions of vertical moisture flux resulted. HELP also over-estimated the moisture flux at the bottom of the landfill in all cases simulated, but excesses were much less and did not vary by as great a margin.

---

[1] Ph. D. Candidate, Civil & Envir. Engrg. Dept., Univ. of California, Davis, CA 95616

[2] Professor, Civil & Envir. Engrg. Dept., Univ. of California, Davis, CA 95616

Understanding these limitations can prevent misapplication of the HELP model, permit its confident use where conditions indicate, and demonstrate where more demanding analysis maybe required.

Introduction

An Environmental Protection Agency (EPA) model Hydrologic Evaluation of Landfill Performance (HELP), Version 3, was tested to evaluate its simulation of vertical moisture transport (Schroeder et al., 1984a, 1984b, 1992, 1994a, 1994b). Comparison of an earlier version of HELP with a one-dimensional finite difference model (UNSAT1D) was performed for the Electrical Power Research Institute (EPRI) by Batelle, Pacific Northwest Laboratories (EPRI 1984). The EPRI report raised concerns about the dependability in simulation of vertical transport using HELP. It found that the earlier version of HELP over-estimated downward moisture flux in a landfill, both through the barrier layer and landfill bottom, under both arid and semi-arid climates. Conclusions of the EPRI report also incorporated differences in the way the two models calculated runoff and evapotranspiration to obtain the net infiltration into the surface. Since UNSAT1D contained no vegetative growth model, only unvegetated surfaces were considered in the EPRI report.

This report compared a beta release of Version 3 of HELP (the cross section of cases were subsequently compared with the release version) with a two-dimensional, finite element, unsaturated ground water flow model, RMA42. RMA42 was developed by Resource Management Associates and has been expanded at University of California, Davis (King et al., 1978). RMA42 (and UNSAT1D) solve the Richards equation, the physical equation of unsaturated ground water flow that accounts for gravity and capillary forces (Richards, 1931). RMA42's solution included unsaturated hydraulic conductivity calculations from a relationship by van Genuchten (Sisson et al., 1980). HELP is a simplified water budget model using Darcy's Law for vertical flow with empirical calculations of unsaturated hydraulic conductivity by the Campbell equation based on the Brooks-Corey relationship (Brooks and Corey, 1964). Extracting the net infiltration values calculated by HELP and applying them to RMA42 as boundary conditions isolated the vertical transport algorithms to provide a direct comparison of the different vertical transport methods simulated.

Results with unvegetated soil surface demonstrated that HELP simulated vertical water fluxes into the waste layer under humid climate conditions reasonably well. As climate conditions became more arid, empirical assumptions used in HELP increasingly limited its ability to predict reasonable design values of vertical transport within a landfill. Results produced with 300% increase in clay barrier layer thicknesses indicated over-predictions are still evident, but that the percentage change is comparable to the Richards equation solution. Simulations performed with vegetated surfaces produced predictions with equal or greater differences than simulations with bare surfaces.

Descriptions of Models

HELP Version 1

HELP Version 1 (Schroeder et al., 1984a, 1984b) was developed to "facilitate rapid, economical estimation of the amount of surface runoff, subsurface drainage, and leachate...". Version 1 advanced the Hydrologic Simulation Model for Estimating Percolation at Solid Waste Disposal Sites (HSSWDS), the first model developed for predicting percolation through landfills. HELP and HSSWDS were sponsored by EPA and developed by the Waterways Experiment Station, Corp of Engineers (WES). Version 1 of HELP incorporated most of the runoff, evaporation, and transpiration routines of the Chemical Runoff and Erosion from Agriculture Management Systems (CREAMS) model from the U. S. Department of Agriculture (USDA). Runoff and infiltration calculations are primarily from the Hydrology Section of the National Engineering Handbook, using the Soil Conservation Service (SCS) runoff curve number method. HSSWDS modeled only the cover layer, while Version 1 included lateral drainage and liner leakage and simulated the entire landfill.

HELP Version 2

Version 2 of HELP (Schroeder 1992) added a synthetic weather generator (WGEN) developed by the USDA Agriculture Research Service. WGEN can simulate up to twenty years of climate input. In addition, both a five year (1974-78) climatological database and manual input option were retained. HELP uses WGEN to calculate daily values of the maximum temperature, minimum temperature, and solar radiation values, for any climate input option chosen. A vegetative growth model from the Simulator for Water Resources in Rural Basins (SWRRB) is used to calculate leaf area indices. Version 2 made refinements to the unsaturated hydraulic conductivity and the lateral drainage flow algorithms, and improvements to the interface to ease use of the model.

HELP Version 3

Version 3 of HELP (Schroeder et al., 1994a, 1994b) contains additional refinements to the interface, including on-line help screens, making it still easier to implement, and modifications to many transport algorithms to improve the predictions. An energy-based model now performs snow melt calculations, and a modified Penman model using a concept by Ritchie calculates evapotranspiration (Ritchie, 1972). Version 3 increases the number of layers that can be modeled and the available default definitions for these layers. At present 42 default layer types are available including geomembranes, geosynthetic drainage nets, and compacted soils. Leakage through membranes and recirculation of leachate is included. Incorporation of the effects of slopes and lengths of surfaces and of frozen soil surfaces improves runoff calculations. It modifies unsaturated vertical moisture flux to improve moisture storage calculations.

Version 3 remains a quasi two-dimensional (the lateral drainage element can be used as needed) deterministic water budget, utilizing a variety of physics-based calculations, empirical estimates, and simplifying approximations. Layer types still define the vertical profile of the system. There are four different layer types available: vertical percolation layers, lateral drainage layers, barrier soil layers, and geomembrane layers. Vertical drainage is still governed by Darcy's law:

$$\text{where ...} \qquad q = K_z(\theta)\frac{dh}{dl}. \qquad (1)$$

The discharge per unit time per unit area normal to the direction of flow, q, is proportional to the unsaturated hydraulic conductivity, $K(\theta)$ a function of moisture content, times the hydraulic gradient given by the change in piezometric head, h, divided by the distance, l, of the flow. Piezometric head, the sum of pressure and elevation, is assumed constant over each vertical percolation layer defined.

Unsaturated Ground Water Flow Model, RMA42

The RMA42 code is a two-dimensional finite element solver of the Richards equation for two-dimensional, unsaturated ground water flow (Bear 1972). The Richards equation, defining unsaturated ground water flow, in two-dimensional form:

$$\text{where ...} \qquad \frac{\partial}{\partial x}\left[K_x(\psi)\frac{\partial\psi}{\partial x}\right] + \frac{\partial}{\partial z}\left[K_z(\psi)\left[\frac{\partial\psi}{\partial z}+1\right]\right] = \Theta(\psi)\frac{\partial\psi}{\partial t} \qquad (2)$$

directly includes both gravity and capillary forces, so water is able to be moved up or down, continuously over the entire domain, depending on the magnitudes of the two opposing driving forces (Freeze and Cherry 1979). Piezometric head has been replaced by pressure head plus elevation, $h = \psi + z$, and q must be considered in terms of moisture content, $\theta(\psi)$, which is also a function of pressure head.

A finite element system is approximated as a continuous series of discrete elements with approximation functions defined in the distribution for each element. The Galerkin method of weighted residuals was used to minimize errors in an integral sense over all the elements. This produced a finite set of equations describing the system. The Richards equation is non-linear in both hydraulic conductivity, $K(\psi)$, and moisture content, $\theta(\psi)$, so a Newton-Raphson iteration scheme derived a set of linear equations.

Application of Models

To make a direct comparison with the EPRI results, the models were applied to simulate a simple landfill with 61 cm (2 ft) of cover soil over a 30.5 cm (1 ft), or 91.4 cm (3 ft), clay barrier layer and a 914.4 cm (30 ft) waste layer profile. For the HELP model, the layers must be identified by drainage type (vertical, lateral, and barrier) with material characteristics (porosity, field capacity, wilting point, and hydraulic conductivity) and initial moisture conditions specified. For RMA42 every element defining each layer required specification of material characteristics as well as capillary pressure to specify initial moisture content.

Top soil and clay barrier layers were typical of those used in municipal solid waste landfills. Material characteristics of the waste layer were chosen to be representative of incinerator ash in a hazardous waste landfill. Cincinnati, OH, Brownsville, TX, and Phoenix, AZ represented three climatic test cases of humid, semi-arid, and arid conditions respectively. All simulations ran over a two year period (1974-75) using the historical climatological database in HELP, (Figure 1). The two years chosen are sufficiently close to historical averages for the regions to be representative. HELP simulated vegetative growth, where utilized.

For each case the HELP code and the Richards equation solution were simulated with identical initial and boundary conditions. Initial moisture content was set at field capacity in each layer except for the evaporative zone (cover layer), which was set to the average of field capacity and wilting point. This is a default option for the HELP model. For RMA42, and UNSAT1D, capillary pressures defined equivalent soil moisture contents. Values of runoff, evapotranspiration, and surface storage combine to produce the net surface infiltration value predicted in each HELP simulation. These net infiltration values were the boundary conditions in the comparable RMA42 simulation.

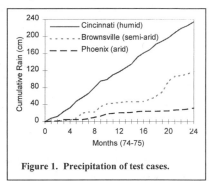

Figure 1. Precipitation of test cases.

Using the same net infiltration values in both models served to isolate and provide a direct comparison of vertical moisture transport calculations. Differences could be expected since RMA42 used a solution of the Richards equation accounting for capillary forces not directly considered in the empirical HELP model. HELP can only remove moisture from the specified evaporative zone, which cannot extend below the top of the barrier layer. Evaporative zone depth must be specified by the user for bare soil or vegetative growth, HELP includes recommended values.

Each model also used a slightly different estimator for unsaturated hydraulic conductivity. We examined differences in the solution of values of unsaturated hydraulic conductivity and determined them to be insignificant. This alleviated the necessity to modify the unsaturated hydraulic conductivity code in RMA42 to duplicate that of HELP.

In all figures and tables in this report, net infiltration represented the moisture flux through the surface, barrier flux represented the flux through the bottom of the clay barrier layer, and bottom flux represented drainage through the bottom of the waste layer. Likewise, in all graphs the fluxes, i.e. cm/day, are time integrated, or cumulative, and given in centimeters of moisture.

The work that is presented here was completed in separate stages. First, we simulated the exact problem performed in the EPRI report using HELP Version 1 and RMA42. The case considered first the thin, 30.5 cm barrier layer and then the thick 91.4 cm, barrier layer without surface vegetation. Results helped to confirm the validity of RMA42 output by direct comparison with UNSAT1D. HELP simulations used a 15 cm evaporative zone depth, recommended at that time for bare soil. Second, the identical simulations were performed using the HELP Beta Version 3 and RMA42. Third, HELP Beta Version 3 and RMA42 simulations were repeated using the current variable evaporative zone depths recommended by HELP. Fourth, the vegetative growth model of HELP was utilized to examine the effects of surface vegetation. Bottom fluxes were examined and comparisons made for all simulations. The final effort used the release version of HELP and investigated revisions made from the beta version used in the first three parts.

Results

EPRI Report - UNSAT1D versus HELP Version 1

The EPRI report concludes, and duplicate work with RMA42 confirmed, that the earlier version of HELP provides reasonable estimates for humid climate conditions compared to a Richards equation solution, but over-predicts downward moisture transport for both semi-arid and arid climates (Tables 1 & 2). EPRI concludes that model comparisons indicate the amount of over-estimation increases as the climate becomes more arid.

For all data reported by EPRI, HELP predicts greater runoff, while UNSAT1D predicts greater evapotranspiration (in this case only evaporation since UNSAT1D has no transpiration algorithms so bare soil is modeled). UNSAT1D uses a simplified runoff calculation that the EPRI report concludes over-estimates infiltration and consequently the soil contains more moisture to evaporate.

During all 2 year simulations, HELP predicts increases in the volume of water stored in the landfill for all climates, while UNSAT1D shows increases in storage only for humid climates. Under semi-arid and arid conditions the UNSAT1D code draws moisture up through the landfill so it becomes available to be evaporated. HELP cannot raise moisture and evaporate it once the moisture is transported below the evaporative zone.

UNSAT1D predicts a nearly constant moisture flux through the bottom of the waste layer. This result is due to a fixed gradient boundary condition at the bottom and the effects of percolation through the cover not reaching the bottom over the time of the simulation. HELP demonstrates that the bottom flux is dependent on the flux through the barrier layer. Bottom fluxes diminish as the barrier layer fluxes decrease, until bottom flux becomes nearly zero when zero barrier flux is predicted.

HELP Beta 3 Version versus Version 1

Comparing the Beta Version 3 of HELP with Version 1, Version 3 of HELP predicted less runoff for all simulations, both barrier layer thicknesses, and for identical and revised evaporative zone depths and vegetated surfaces. Estimations of evapotranspiration generally increased except for the thick barrier layer under arid climate with a 15 cm evaporative zone (current recommendation is 45.7 cm).

Version 3 gave greater estimates for barrier layer flux for all simulations while it lowered the moisture storage for the system for all but the no longer recommended 15 cm evaporative zone depths. This resulted in generally greater moisture content in the waste layer itself.

HELP Beta 3 Version versus RMA42 - 15 cm evaporative zone depth

Simulations identical to the EPRI report were performed with HELP Beta Version 3 and RMA42. They utilized a 15 cm evaporative zone depth, recommended in the earlier version of HELP for bare soil, for both the thin and thick barrier layers. Although not necessarily expected to produce new results, this provided additional verification of RMA42 with the EPRI results, exposed the issue of the UNSAT1D model's runoff and infiltration algorithms, and provided the comparison of the new release of HELP with the earlier version described above.

Simulations on 30.5 cm barrier layer with unvegetated surface supported the conclusions of the EPRI report (Figures 2, 3, & 4). Under humid conditions, although HELP demonstrated insensitivity to infiltration variations, HELP estimations compared favorably with predictions of RMA42 (Figure 2). Otherwise, HELP generally predicted greater flux than RMA42 through the barrier layer by increasingly greater amounts as the climate became more arid.

For the semi-arid climate, net infiltration (Figure 3) was more cyclic than under humid conditions, and further demonstrated that HELP remained insensitive to cyclic boundary conditions. Without regard to net negative infiltration during the early months of the simulation, HELP predicted a continual, linear increase in barrier flux. The Richards solution of RMA42 demonstrated both a reasonable lag and response to the boundary conditions. HELP estimated barrier flux values in excess of 18% above RMA42.

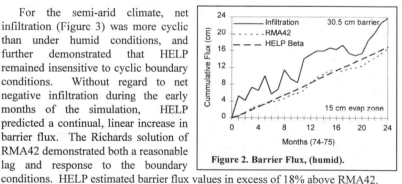

Figure 2. Barrier Flux, (humid).

Arid climate simulation (Figure 4) showed HELP predicted greater downward flux through the barrier layer in increasingly larger amounts as the climate became more arid, now 45% above RMA42. HELP's linear prediction of downward flux

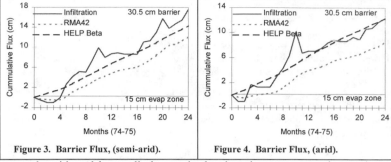

Figure 3. Barrier Flux, (semi-arid).   Figure 4. Barrier Flux, (arid).

was only mitigated by small changes in the slope in response to changes in net infiltration. RMA42 more obviously registered changes throughout the simulation reflecting infiltration adjustments.

For a 91.4 cm barrier layer, unvegetated surface the simulations were equivalent to the examples of the EPRI report with the thicker barrier (Figures 5, 6, & 7). RMA42 results again supported the original conclusions of the EPRI report.

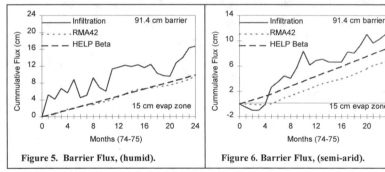

Figure 5. Barrier Flux, (humid).   Figure 6. Barrier Flux, (semi-arid).

As expected, both models predicted lower net surface infiltration and less downward barrier layer flux with the thicker barrier layer. As a trend, HELP still over-predicted barrier flux in greater amounts as the climate became more arid (3%, 26%, & 40%). While the reduction in barrier layer flux due the thicker barrier layer varied with climate, HELP and RMA42 predicted nearly identical percentage reductions for the same climate conditions.

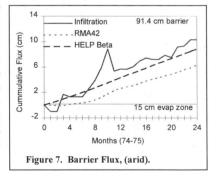

Figure 7. Barrier Flux, (arid).

It is clear that a 300% thicker barrier layer, for a fixed 15 cm evaporative zone depth, only reduced net surface infiltration and barrier layer fluxes. It did not change the trend of either value or their relationship to one another. The thicker barrier layer retarded downward moisture transport giving HELP additional time to remove increased moisture through evaporation.

<u>HELP Beta 3 Version versus RMA42 - Variable Evaporative Zone Depth</u>

HELP, in its present form, suggests evaporative zone depths for bare soil conditions that vary with the climate condition. Evaporative zone depths now recommended for bare soil in humid, semi-arid, and arid climates are 22.9 cm, 30.5 cm, and 45.7 cm respectively. Examination of these simulations provided another step to assess the ongoing modifications to HELP.

For the 30.5 cm barrier layer, as the evaporative zone became deeper, the net surface infiltration decreased in proportion (Figures 8, 9, 10).

Under humid and semi-arid climate conditions, the increased evaporative zone depths provided a greater volume from which HELP could remove moisture from the landfill. Removal of more moisture resulted in subsequent greater net infiltration amounts, higher precipitation followed the surface drying permitting greater infiltration. While overall this resulted in less net surface infiltration, it also resulted in increased cycling of the net surface infiltration.

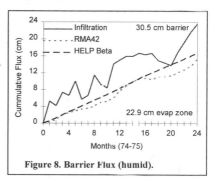

Figure 8. Barrier Flux (humid).

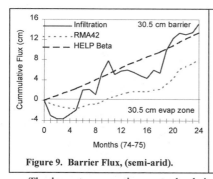

Figure 9. Barrier Flux, (semi-arid).

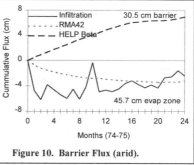

Figure 10. Barrier Flux (arid).

The largest evaporative zone depth increase, for the arid climate (Figure 10), resulted in negative net surface infiltration (net removal of moisture from the landfill). Accompanying reductions in the barrier layer fluxes predicted were not as great as the reductions of the surface infiltration. This resulted in a greater disparity in the barrier flux values predicted by HELP versus RMA42 for each climate.

Arid climate simulations (Figure 10) showed HELP continued to predict positive downward flux through the barrier even when the net infiltration was upward out of the landfill. HELP's linear prediction of downward flux continued until about month 16 when the evaporative zone depth dried out and the curve leveled off. HELP then transported more moisture downward from the precipitation events of the last 3 months. RMA42 more properly predicted the continued removal of moisture up through the barrier layer throughout the simulation with slope changes responding to infiltration variations.

Results of revised evaporative zone depths applied to the 91.4 cm barrier case (Figures 11, 12, & 13) supported the conclusions found in barrier layer thickness comparisons of 15 cm evaporative zone depths.

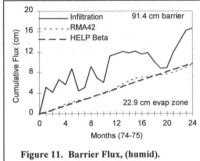

Figure 11. Barrier Flux, (humid).

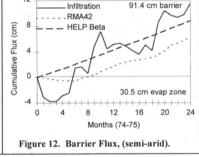

Figure 12. Barrier Flux, (semi-arid).

Again, both models predicted lower net surface infiltration and less downward barrier layer flux with the thicker barrier layer. For the humid and semi-arid climates, HELP and RMA42 predicted nearly identical percentage reductions in barrier layer flux. While the percentage reduction of barrier layer flux did not hold for the arid climate, the RMA42 value is nearly zero so the percentage is very sensitive to small changes.

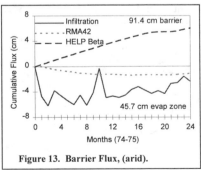

Figure 13. Barrier Flux, (arid).

The 300% thicker barrier layer again only resulted in reduced net surface infiltration and barrier layer fluxes. It did not change the trend of either value or their relationship to one another.

HELP Beta 3 Version versus RMA42 - Vegetated Surface

While not considered in the EPRI study, since UNSAT1D did not contain the required algorithms, simulations of the same landfill structure with a vegetated surface were performed. With vegetation HELP recommends adjusted evaporative zone depths and calculated leaf area indices that contributed to revised evapotranspiration predictions. Revised values of net infiltration were again applied to RMA42 as a boundary condition.

For the 30.5 cm barrier, the humid climate with surface vegetation produced cyclic values of infiltration to which HELP responded only by slope changes of the still increasing barrier flux, (Figure 14). Barrier flux predicted by RMA42 closely reflected the negative and positive variations of the infiltration process.

Semi-arid conditions with vegetation (Figure 15) reduced the cyclic infiltration of the bare surface to primarily negative values. HELP continued its linear

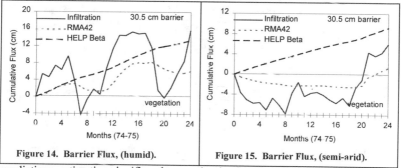

| Figure 14. Barrier Flux, (humid). | Figure 15. Barrier Flux, (semi-arid). |

predictions, estimating significantly more barrier flux than RMA42, and calculating more barrier flux than net surface infiltration. Previously barrier flux only exceeded net infiltration for arid climates, but the vegetation removed enough additional moisture to lower the net infiltration while the HELP algorithm continues to transport moisture below the evaporative zone downward.

For the arid climate with vegetation, HELP ceased prediction of additional barrier flux after 5 months (Figure 16). Drainage of initial moisture content from the evaporative zone and rainfall early in the simulation ceased and, without additional infiltration, the barrier flux after 5 months stopped increasing. RMA42 code predicted a gradually tapering rate of flux up through the barrier layer. This upward

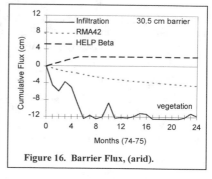

Figure 16. Barrier Flux, (arid).

flux, caused by capillary forces, is not accounted for in the HELP model. HELP only has provisions to remove moisture from the soil evaporative zone depth defined by the user. Once moisture is transported below this level it can only be transported downward or held.

Bottom Flux Comparisons

Flux through the bottom of the waste layer demonstrated the attenuation expected through the depth of the landfill (Figures 17 & 18). Here data are presented to represent a bottom flux comparison of climate extremes with the thin barrier layer thickness for 15 cm evaporation zone depth and vegetated simulations. The humid climate with 15 cm evaporation zone depth and the arid vegetated surface simulations represented the greatest extreme found in net surface infiltration (23.9 cm versus -11.8 cm), in barrier layer flux (17 cm versus 2.2 cm), and in the bottom flux (3.9 cm versus 3.1 cm).

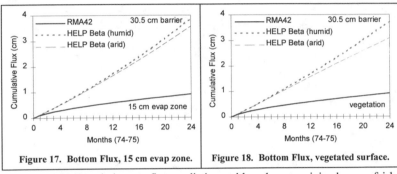

Figure 17. Bottom Flux, 15 cm evap zone.   Figure 18. Bottom Flux, vegetated surface.

The RMA42 code bottom flux predictions, although very minimal, were fairly constant regardless of climate since bottom flux is a function of the boundary condition applied to the bottom of the landfill profile. The lower boundary condition was set at a sufficient gradient 30.5 cm below the bottom of the waste layer, allowing free drainage of the waste layer.

Bottom fluxes of the two extreme climate conditions for the 15 cm evaporative zone depth (Figure 17) demonstrated only 5% difference even though the barrier layer fluxes varied by over 26% (Figures 3 & 5). Similarly, for the vegetated surfaces, the bottom fluxes varied only 16% while the barrier fluxes varied by over 450% (Figures 14 & 16).

HELP Release Version 3

Following completion of the work presented above, HELP Version 3 was officially released. Comparative simulations were performed with the newly released version to determine results of any subsequent changes in the code (Figures 19, 20, 21, 22, 23, & 24). Results presented above demonstrated that barrier flux trends predicted by RMA42 had a close relationship to net infiltration and prevented the necessity of duplicating the transport with the RMA42 code.

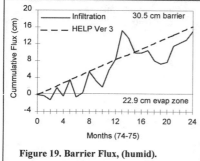

Figure 19. Barrier Flux, (humid).

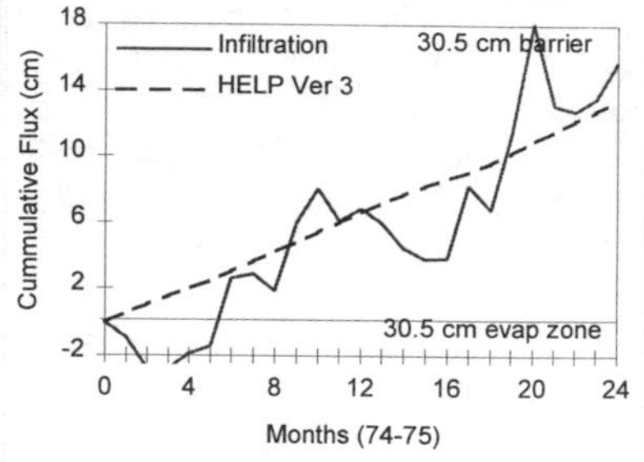

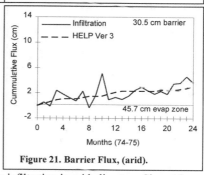

Figure 20. Barrier Flux, (semi-arid).

There are clearly some significant computational changes in the release version that are not explicitly described its the accompanying documentation. Runoff values predicted increase with increasing humid climates, while evapotranspiration increases in humid climates, decreases slightly in semi-arid climates, and increases even more in arid climates. This results in substantially less net infiltration in humid climates, almost no change in

Figure 21. Barrier Flux, (arid).

semi-arid climates, and significantly more infiltration in arid climates. Since 15 cm evaporative zone depths (Tables 1 & 2) are no longer valid recommendations, the data were ignored for analysis. The release version predicted increased runoff and evapotranspiration for simulations with the humid climate. This produced a 36-60% decease in net infiltration, and contributed to similar reductions in landfill moisture change (Figures 19 & 22).

In the arid climate, where no runoff was predicted for the beta or release versions, evapotranspiration decreased eliminating the negative net infiltration rates of the beta version (Figures 21 & 24). In the most extreme case, thin barrier layer with vegetation where net infiltration was only 0.1 cm, HELP predicted smaller negative values of infiltration over the 2 year simulation

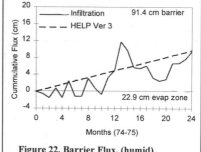

Figure 22. Barrier Flux, (humid).

and less often than either of the arid simulations shown. The semi-arid climate demonstrated mixed changes (Figures 20 & 23). For the unvegetated thick and thin barrier layers the runoff increased while the evapotranspiration decreased, which left the net infiltration mostly unchanged. However the vegetated thin barrier layer demonstrated a small decrease in runoff and an increase in evapotranspiration, which increased the net infiltration. Bottom flux results from the release version of HELP were no longer consistently high, but remained at the previous levels for lower values of barrier layer flux.

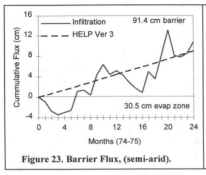

Figure 23. Barrier Flux, (semi-arid).

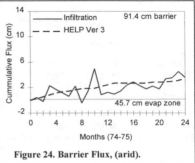

Figure 24. Barrier Flux, (arid).

Summary and Conclusions

An historical analysis was performed on the EPA HELP model that reviewed the origins and changes in the model, and examined the results that it produced. There is no question that an empirical water budget model is much easier to solve computationally. HELP, using empirical algorithms for vertical moisture transport, retained some inherent deficiencies in spite of continual improvements. The user must determine if these shortcomings will effect the design criteria.

Except where infiltration showed a significant spike near the end of the simulation, examination of the HELP results indicated that the barrier layer flux estimates nearly equaled or exceeded the net surface infiltration over the 2 year simulation period for all climates and barrier layer thicknesses. Interpolation of the RMA42 Richards equation solution of similar infiltration patterns indicated a more logical response of the barrier layer flux, where barrier flux lagged behind the net surface infiltration. The release version of HELP produced predictions of net surface infiltration in arid climates more closely to those of conventional wisdom, but barrier layer fluxes are still greater than would be produced with a Richards equation solution.

Percentage decreases of barrier flux estimates for HELP and RMA42 in humid and semi-arid climates were very consistent. This indicated HELP should give usable percentage change estimations for barrier layer thickness changes, even if absolute values are questionable as design values. The release version upheld these findings, but still did not improve the results in arid climates. Bottom flux predictions by HELP also remain problematic. While the beta version consistently estimated higher bottom flux, the release version of HELP still predicted higher bottom flux for the lower values of barrier layer flux predicted. This error prevents a similar use of the program for percentage change of comparative bottom flux predictions, and total moisture change, that is possible for barrier layer fluxes. In defense of HELP, bottom fluxes will always be more variable than surface infiltration and barrier flux due to heterogeneities in most landfills. Simulations here focused on a more homogeneous fly ash landfill, rather than reality.

Without specific modification to the empirical HELP code to more closely account for capillary forces and to be able to remove water from below the soil evaporative zone, HELP will continue to over-predict downward vertical moisture fluxes. This accelerated downward movement of moisture will cause associated difficulties in the infiltration and runoff calibrations, since both are dependent functions of surface soil moisture content. Attempting to calibrate for any of these three variables will adversely affect the other two unless the downward movement is more physically based. With a desire to better control gas production bringing more interest in the water balance within the landfill, a more accurate definition of vertical drainage may be required.

Meanwhile, a complete understanding of these limitations can allow confident use of HELP where conditions permit, and demonstrate where more rigorous analysis is warranted.

Data

Table 1. Comparison of Simulation Results for all 30.5 cm Clay Barrier Layers

Location & Evap Zone	Model Used	Total Precip	Total Runoff	Total ET	Barrier Flux	Bottom Flux	Moisture Change
Humid (Cincinnati, OH)							
15 cm bare	HELP1	234.8	65.8	141.5	13.9	4.3	18
	UNSAT1D	234.8	41.9	161	21	3.1	26.9
15 cm bare	BETA3	234.8	49.1	161.8	17	3.9	20
	RMA42	234.8	49.1	161.8	16.7	0.9	23
	HELP3	234.8	53.1	166.1	16.7	0.83	14.8
22.9 cm bare	BETA3	234.8	47	164.2	16.8	3.9	19.7
	RMA42	234.8	47	164.2	15.8	0.9	22.7
	HELP3	234.8	51.6	168.2	16.3	0.82	14.2
53.3 cm veg	BETA3	234.8	30.8	188.2	13.4	3.7	12.1
	RMA42	234.8	30.8	188.2	6.1	0.9	14.9
	HELP3	234.8	37.2	191.3	11.9	0.80	5.5
Semi-Arid (Brownsville, TX)							
15 cm bare	HELP1	117.8	30.7	74.6	7.5	1.9	10.6
	UNSAT1D	117.8	0	136.6	-5.2	3.1	-21.9
15 cm bare	BETA3	117.8	25.2	74.9	14.3	3.7	13.9
	RMA42	117.8	25.2	74.9	12.1	0.9	16.7
	HELP3	117.8	25.2	75.8	15.0	0.82	16.0
30.5 cm bare	BETA3	117.8	14.8	87.8	13.4	3.6	11.5
	RMA42	117.8	14.8	87.8	8.0	0.9	14.2
	HELP3	117.8	16.4	85.6	13.5	0.79	14.7
58.4 cm veg	BETA3	117.8	9.7	102	9.3	3.4	2.7
	RMA42	117.8	9.7	102	1.4	0.9	5.1
	HELP3	117.8	8.98	101	3.84	3.27	4.6
Arid (Phoenix, AZ)							
15 cm bare	HELP1	32.2	0.6	30.9	0	0.1	0.7
	UNSAT1D	32.2	0	54.5	-5.8	3.1	-25.4
15 cm bare	BETA3	32.2	0	20	12.3	3.6	8.7
	RMA42	32.2	0	20	8.5	0.9	11.4
	HELP3	32.2	0	20.7	12.7	0.74	10.8
45.7 cm bare	BETA3	32.2	0	34.6	7	3.4	-5.7
	RMA42	32.2	0	34.6	-1.7	0.9	-3.3
	HELP3	32.2	0	28.5	3.09	2.21	1.5
60.96 cm veg	BETA3	32.2	0	44	2.2	3.1	-14.9
	RMA42	32.2	0	44	-4.5	0.9	-12.7
	HELP3	32.2	0	32.1	0	3.08	-2.98

(all values in centimeters, shaded areas reported in EPRI results)

Table 2. Comparison of Simulation Results for all 91.4 cm Clay Barrier Layers

Location & Evap Zone	Model Used	Total Precip	Total Runoff	Total ET	Barrier Flux	Bottom Flux	Moisture Change
Humid (Cincinnati, OH)							
15 cm bare	HELP1	234.8	73.7	141.1	5.7	2.2	18
	UNSAT1D	234.8	51.3	161	11.9	3.2	17.4
15 cm bare	BETA3	234.8	54.9	163.2	9.9	3.4	13.3
	RMA42	234.8	54.9	163.2	9.6	1.1	15.6
	HELP3	234.8	57.9	167.5	9.84	0.84	8.56
22.9 cm bare	BETA3	234.8	53.9	164.3	9.9	3.4	13.3
	RMA42	234.8	53.9	164.3	9.8	0.7	15.9
	HELP3	234.8	57.6	167.9	9.73	0.82	8.48
Semi-Arid (Brownsville, TX)							
15 cm bare	HELP1	117.8	32.7	75	3.8	1.1	9.1
	UNSAT1D	117.8	0	139.3	-4.1	3.2	-24.2
15 cm bare	BETA3	117.8	27.1	77.8	9.1	3.4	9.5
	RMA42	117.8	27.1	77.8	7.2	1.1	11.8
	HELP3	117.8	28.1	78.8	9.45	0.83	10.1
30.5 cm bare	BETA3	117.8	17.2	89	8.9	3.4	8.2
	RMA42	117.8	17.2	89	6.3	0.8	10.7
	HELP3	117.8	18.4	88.7	9.01	0.79	9.91
Arid (Phoenix, AZ)							
15 cm bare	HELP1	32.2	0.6	30.9	0	0.1	0.8
	UNSAT1D	32.2	0	58	-4.3	3.1	-28.9
15 cm bare	BETA3	32.2	0	21.9	8.8	3.4	6.9
	RMA42	32.2	0	21.9	6.3	1.1	9.2
	HELP3	32.2	0	23.2	9.0	0.75	8.25
45.7 cm bare	BETA3	32.2	0	35.5	6.1	3.3	-6.5
	RMA42	32.2	0	35.5	-1.1	0.8	-4.0
	HELP3	32.2	0	28.6	3.42	1.84	1.76

(all values in centimeters, shaded areas reported in EPRI results)

Nomenclature

ψ,	pressure head (negative capillary pressure)
$K_x(\Psi), K_z(\Psi)$,	non-linear hydraulic conductivity
$\theta(\Psi)$,	non-linear moisture content
x, z,	horizontal and vertical Cartesian coordinates
t,	time

References

Bear, Jacob, *Dynamics of Fluids in Porous Media* (1972), Dover Publications, Inc., New York, New York

Brooks, R. H. and Corey, A. T. (1964), *Hydraulic properties of porous media*, Hydrology Paper No. 3, Colorado State University, Fort Collins, CO. 27pp.

Comparison of Two Groundwater Models - UNSAT1D and HELP (1984), Palo Alto, California, Electric Power and Research Institute, CS-3659.

Freeze, R.A., and Cherry, J.A. (1979), *Groundwater*, Prentice-Hall, Inc., Englewood Cliffs, New Jersey.

King, Ian P., Norton, W. R., and McLaughlin, D. B., *Model for Simulation of Natural Moisture Movement through the Hanford Vadose Zone* (1978), Richland, WA, report submitted to Rockwell International Corporation Atomics International Division by Resource Management Associates, RMA 7080.

Richards, L.A. (1931) "Capillary conduction of liquids through porous mediums", *Physics*, 1, pp. 318-333.

Ritchie. J. T. (1972) "A model for predicting evaporation from a row crop with incomplete cover", *Water Resources Research* 8(5), 1204-1213.

Schroeder, P.R., Morgan, J.M., Waliski, T.M., and Gibson, A.C. (1984), *The Hydrologic Evaluation of Landfill Performance (HELP) Model. Vol I. User's Guide for Version I*. PB85-100840. U.S. Environmental Protection Agency, Office of Solid Waste and Emergency Response, Washington, D.C.

Schroeder, P.R., Morgan, J.M., Waliski, T.M., and Gibson, A.C. (1984), *The Hydrologic Evaluation of Landfill Performance (HELP) Model. Vol II. Documentation for Version I*, PB85-100832. U.S. Environmental Protection Agency, Office of Solid Waste and Emergency Response, Washington, D.C.

Schroeder, P. R. (1992), *Interim Guide for HELP Version 2 for Experienced Users*. personal correspondence.

Schroeder, P.R., Lloyd, C.M., and Zappi, P.A. (1994), *The Hydrologic Evaluation of Landfill Performance (HELP) Model User's Guide for Version 3*, EPA/600/R-94/168a, U.S. Environmental Protection Agency Risk Reduction Engineering Laboratory, Cincinnati, OH.

Schroeder, P.R., Dozier, T.S., Zappi, P.A., McEnroe, B.M., Sjostrom, J.W., and Peyton, R.L. (1994), *The Hydrologic Evaluation of Landfill Performance (HELP) Model: Engineering Documentation for Version 3*, EPA/600/R-94/168b, U.S. Environmental Protection Agency Risk Reduction Engineering Laboratory, Cincinnati, OH.

Sisson, J.B., Furgeson, A.H., and M. Th. van Genuchten (1980), "Simple Methods for Predicting Drainage from Field Plots", *Soil Sci. Soc. Amer. J.*, Vol. 44, pp. 1147-1152.

Design and Construction of Foundations Compatible With Solid Wastes

R. Jeffrey Dunn,[1] Member, ASCE

Abstract

Closed landfills are increasingly being developed for a variety of purposes. Development often includes construction of structures over landfill waste materials, usually municipal solid waste (MSW). A variety of processes, notably decomposition of the waste, often results in sizable settlements which limit the use of shallow foundations to small lightly loaded structures. Most structures have to be supported on pile foundations driven into competent supporting materials below the landfilled wastes. Foundations can and have been successfully designed and constructed below landfills. Successful foundation design requires careful consideration and evaluation of a variety of factors. Analysis of landfill settlement, bearing capacity of shallow foundations, and capacity of deep foundations, downdrag due to waste settlement are all issues which must be evaluated as part of foundation design for landfill development. Equally important are issues of environmental protection and ensuring that foundations do not lead to environmental degradation by breaching landfill containment systems. Additionally, construction must be carefully planned to deal with the characteristics of wastes and must always thoroughly consider health and safety issues associated with landfills. Design of foundation at landfills requires careful and detailed design, failure to properly consider the many factors associated with foundations over or through wastes can result in poor performance. However, foundations have been successfully built at landfills and are performing very well.

Introduction

During the last fifteen to twenty years there has been an increasing use of

---

[1] Principal, GeoSyntec Consultants, 1600 Riviera Avenue, Suite 420, Walnut Creek, California 94596

solid and industrial waste landfills for a variety of post-closure land uses. Projects have ranged from parks or dedicated open space to commercial and industrial development projects of a wide variety. In some cases, residential development has also occurred on closed landfill facilities, although these projects tend to be less common than other types of development. This is owed to generally negative public perception and increased liability exposure regarding the safety of waste disposal facilities, particularly those in close proximity to housing. A major driving force behind development of closed landfills, previously considered to be derelict land, has been that land in urban areas has become increasingly scarce and expensive. This has made landfills and other marginal lands, which generally cost more than more conventional parcels to develop and maintain, economically viable and, in some cases, very attractive for development.

A number of successful landfill post-closure development projects currently exist in the United States. In addition, there are a few cases where developments have not been so successful. In a few rare cases, problems were large enough that projects have had to be abandoned. In virtually all of these cases that the author is aware of, problems encountered could be traced directly to design and construction which failed to deal effectively with the specific characteristics that landfills exhibit. Thus unsatisfactory performance in these cases has not been surprising. However, these situations are no different than other unsuccessful projects on non-landfill properties where design has failed to consider specific site characteristics. On occasion developments have unknowingly been constructed on waste deposits which are discovered either during or after construction. This situation provides a somewhat different range of problems than post-closure development of a known landfill and thus is not the direct topic of this paper. This type of situation should not occur if a suitable investigation of the property history and subsurface conditions is completed.

A wide variety of structures requiring suitable foundation support may be built as part of a post-closure project. Examples of possible structures include:

- park facilities - sports, restrooms, community centers, and marina facilities;

- commercial developments - retail stores, warehouses, office buildings, manufacturing facilities;

- residential - single-family homes, townhomes, apartments; and

- closure and remediation facilities - maintenance buildings, ground water or leachate treatment plants, and landfill gas treatment plants or flares.

Suitable foundation design for support of structures must include:

- analysis of generally large total and differential settlements and design of foundations and structures compatible with settlements;

- consideration of foundation constructibility and special construction requirements including health and safety of construction personnel; and

- environmental considerations which include the possible impacts of foundations on air and water quality and the impacts of the landfill environment on foundations, mainly through corrosion.

Settlement

As is generally the case with structures built on conventional sites, the magnitudes of total and differential settlements are nearly always the controlling issue in the selection of foundations for structures built on closed landfills. As is discussed in this section, total and differential settlements of landfilled municipal solid waste (MSW) can be sizable.

Settlements of landfilled MSW result from one or more of four basic mechanisms (Oweis and Khera, 1990 and Sharma and Lewis, 1994):

- mechanical or physical compression as a result of the reduction of void spaces or compression of loose material from the MSW self-weight or overlying loads such as landfill covers, engineered fills, and stockpiled soils or waste ;

- raveling or occasional movement of smaller particles in to larger voids which form as the result of collapse, seepage, or vibrations such those induced by an earthquake;

- chemical alterations of MSW by percolating waste liquids or leachate; and

- biological decomposition which tends to be accelerated by high moisture content, warm temperatures, poor MSW compaction, and an increased proportion of organic waste constituents.

Settlements also occur in earthen foundation materials underlying landfills. Settlements in the underlying materials typically are less than those due to landfilled MSW unless the thickness of MSW is small. However, foundation settlement should also be analyzed, particularly for landfill sites underlain by soft clay soils, if new loads are to be applied, or the MSW fill is fairly new so that the materials underlying the foundation are still responding to imposed loads. Conventional techniques should be used for these analyses.

Magnitude and Rate of Settlement. Physical or initial compression of landfilled MSW in response to self-weight or application of fill or surcharge loads typically occurs relatively rapidly. Reviews of landfill performance data indicate that initial compression is essentially complete in 10 to 100 days (Bjarngard and Edgars, 1992 Fassett et al., 1994 and Stulgis et al., 1995). Chemical and biological decomposition processes generally take years to reach completion, with the specific rates varying widely depending upon characteristics of a landfill site and the MSW it contains. Settlements due to raveling are generally difficult to predict and seem to contribute a small percentage to the total landfill settlement as compared to other mechanisms.

Total landfill settlements can be quite large with magnitudes as large as 25% or more of the initial MSW fill thickness. Tchobanoglous et al. (1993) provide general settlement magnitudes suitable for use in very roughly estimating settlements. This information shows that settlements vary significantly depending upon the degree of compaction applied to the MSW at the time of waste placement. Newer, well run landfills apply a high level of compaction to the MSW to maximize disposal capacity which results in lower total settlement of MSW of 25% or less of the original MSW thickness. MSW in older or newer poorly operated landfills typically receives only limited levels of compaction which can increase overall settlements to reported levels as high as 50% of original thickness. It should be emphasized, however, that much of this total settlement occurs rapidly as MSW is placed, so final settlements which might effect development are typically much less than those values. Total settlements also vary with quantities of decomposable materials in MSW. Increasing amounts of inert material tend to decrease settlements to lower magnitudes.

As a direct result of the heterogeneity which is typical of MSW, and also because a landfill may contain variable MSW thicknesses, differential settlements can also be highly variable. Data from a number of landfill sites with refuse thicknesses of approximately 8 to 10 meters, indicated measured differential settlements on the order of 25% over a horizontal distance of 30 meters for time periods of up to 15 years. (Rinne et al., 1994).

Analysis of MSW Settlement. Settlements of solid MSW are not as well understood as those for soils and thus are more difficult to calculate. Stulgis et al. (1995) presented a brief overview of various methods which have been used to analyze the settlement of MSW. Sowers (1968, 1973) first proposed the application of a "soil mechanics" type approach to the analysis of settlements. Others such as Sheurs and Khera (1980) have presented data which show MSW to behave in a manner similar to peat, which is not unexpected in that peat and most MSW is composed of organic material to varying degrees. Edil et al.

(1990) presented data and analysis procedures using a rheological model based upon the Gibson and Lo theory or the power creep law.

Recently, two similar studies collected data from a number of MSW landfills and developed empirical models for the prediction of landfill settlements (Stulgis, 1995). Bjarngard and Edgers (1990) proposed a model based upon behavior shown graphically in Figure 1 which is based upon:

$$\frac{\Delta H}{H} = CR \log \frac{P_O + \Delta P}{P_O} + C_{\alpha(1)} \log \frac{t_{(2)}}{t_{(1)}} + C_{\alpha(2)} \log \frac{t_{(3)}}{t_{(2)}}$$

where:

ΔH = settlement
H = initial thickness of waste layer
$\Delta H/H$ = vertical strain
P_o = initial average vertical effective stress
ΔP = average induced vertical stress increment
$t_{(1)}$ = time (days) for completion of "initial" compression
$t_{(2)}$ = time (days) for completion on "intermediate" compression
$t_{(3)}$ = time (days) for prediction of settlement
CR = compression ratio
$C\alpha_{(1)}$ = intermediate secondary compression index
$C\alpha_{(2)}$ = long term secondary compression index

Fassett et al. (1994) propose a very similar model in which the intermediate and long term compression indices are combined into a single index denoted $C\alpha$ which is applied to calculate compression over a time period between t_i and t_f.

While the use of empirical equations consistent with "soil mechanics" principles seems to represent a reasonable technique for the estimation of landfill settlements, the application of soil mechanics models, even with empirical data represents at best an estimating tool for evaluation of landfill settlements which should be applied carefully and conservatively. At this time, the analysis of MSW in a manner similar to organic soils represents a suitable practice. However, the relative contribution of mechanical compression, thermal effects, and biological decomposition have not been sufficiently ascertained. As is shown by data compiled by Fassett et al. (1994), secondary compression indices can vary considerably in relation to varying MSW compositions, landfill geometries, and operational characteristics. This led Fassett et al. (1994) to further conclude that reported values may only really be useful for application in site specific analyses.

Published data can be used to develop conservative initial settlement

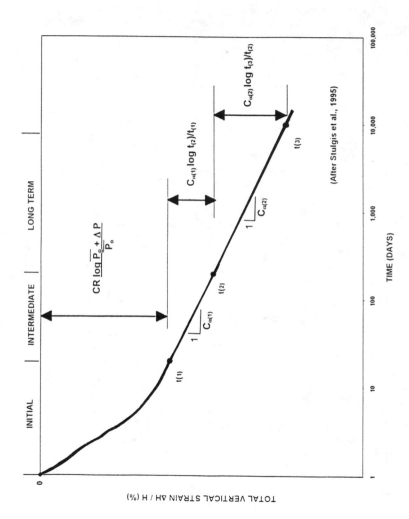

Figure 1 Solid Waste Settlement Model

estimates. However, site specific measurements of settlement versus time are essential to provide a high enough degree of reliability for that analysis and design can be completed with less levels of conservatism. Thus it is strongly recommended that settlement monitoring be implemented during the early stages of a post-closure development project to provide site specific data which is more directly applicable than that found in the literature. If possible, the program should include completion of a test fill program which is monitored for at least a period of 100 days, similar to that described by Stulgis et al. (1995). Alternatively, Stulgis et al. (1995) suggest sampling and analysis of the decomposable fraction of the MSW in place which can then be used to predict volumetric strain based upon the assumption of a relatively constant dry unit weight of the MSW mass which has been observed in several studies (Oweiss and Khera, 1986 and Fassett et al., 1994).

If a test fill is not possible, several other techniques may be implemented which can prove useful to provide site specific data. These techniques do not provide data as reliable as actual monitoring so they should be utilized carefully. The best alternative technique is surveying site settlement at a number of permanent settlement monitoring locations, including, at a minimum, surveying along two or more profiles across the landfill to provide an indication of differential settlements. The survey data can then be plotted versus time on a logarithmic scale which allows calculation of the secondary compression index for the monitoring location. A possible second alternative is comparison of historic topographic maps for the landfill which may show sizable amounts of settlements. This technique is usually limited to sites where settlement has been greater than 0.5 to 1 m, since most topographic maps have a minimum contour interval of 0.6 m (2 ft), and photogrametric and reproduction errors can skew the indication of actual settlements. Additionally, post-closure maintenance activities, such as regrading of landfill covers, can induce errors in the settlements estimated by this technique. However, the method does provide at least a rough indication of settlements.

Another point related to large settlements in landfills and developments which are compatible with settlements, is that construction and post-closure development of a MSW landfill may slow down the rate of MSW decomposition and thus the rate of settlement. This is particularly true at landfills where the cover in place prior to development of the site is not very effective at preventing infiltration from entering the MSW mass. At this type of site development activities usually lead to decreased infiltration whether due to construction of a higher quality cover and/or the decreased site area available for infiltration owing to coverage by structures or paved parking areas. Fortunately, this affect works to reduce settlement rates versus time.

In some structures total settlements up to approximately of 0.1 m (4 in.) can be accommodated by constructing a rigid mat foundation system which can also handle sizable potential differential settlements. In most landfills, unless the MSW thickness is very thin or the MSW has been in place for a sufficient period of time to have undergone most of its decomposition and thus volume reduction, settlements will exceed 0.1 m (4 in.). In some cases, an alternative might be the use of one or more site improvement techniques. Some of these techniques include (Dunn, 1993):

• allowing the MSW to reach an acceptable level of decomposition, either by delaying construction or enhancing decomposition if allowed by environmental regulations;

• supplemental compaction of the MSW, which is usually limited to relatively shallow MSW depths of no more than two or three meters;

• surcharging, with settlement monitoring;

• dynamic compaction; and

• grouting or fly-ash injection.

If site improvement techniques prove to be uneconomical or can not be used to reduce the settlements to tolerable levels, deep foundation systems, usually driven piles must then be utilized.

Foundations

Settlements at landfill sites are nearly always of sufficient magnitude that only lightly loaded structures such as wood frame construction, of two or at most three stories in height, can be supported on shallow foundations. Most other structures require deep foundations.

Shallow Foundations. Shallow foundations at landfills may include the types shown schematically in Figure 2:

• conventional spread footings;

• reinforced concrete mats; and

• grid foundations consisting of column footings tied together with a system of grade beams and usually an integral concrete floor.

Use of conventional spread footings is generally limited to structures which can tolerate differential settlements. Alternatively a less settlement tolerant structure might be constructed with the ability to adjust individual column levels to

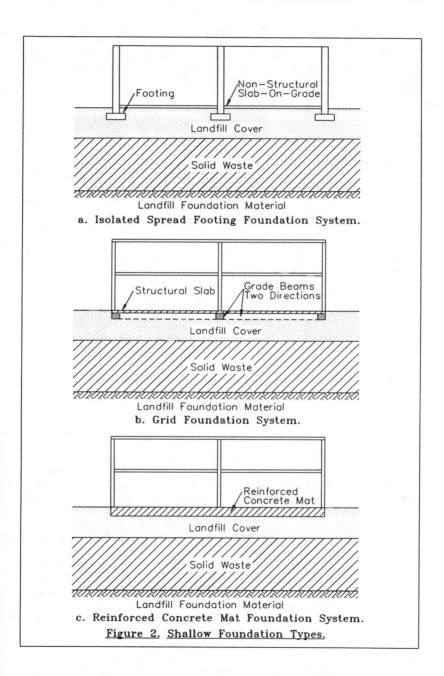

a. **Isolated Spread Footing Foundation System.**

b. **Grid Foundation System.**

c. **Reinforced Concrete Mat Foundation System.**

Figure 2. Shallow Foundation Types.

accommodate settlement by the incorporation of mechanical jacks. However, with the additional cost of the releveling system and cost of maintenance activities, use of a more costly and stronger foundation, which is more settlement tolerant, usually becomes the economical choice. Additional reinforcing steel, beyond what is typically required in a conventional site foundations, is usually added to all of these shallow foundation systems to add stiffness and thus to compensate for the uncertainties of settlement at a landfill.

If shallow foundations are used, conventional bearing capacity analyses can be utilized for design. In most cases, it is desirable to provide a layer of engineered fill sufficiently thick to provide essentially all of the bearing capacity. When this can not be provided, and the MSW will provide some of the bearing capacity, the most difficult aspect of the analysis becomes evaluating the strength of the MSW. Information on the strength properties of MSW is limited, and most of the published data such as that by Jessberger and Kockel (1991), Mitchell and Mitchell (1992), and Howland and Landva (1992), has been published with regard to analysis of slope stability. Fassett et al. (1994) present a good summary of MSW strength properties. As with characteristics of compressibility, it appears that application of soil mechanics principles, in particular Mohr-Coulomb theory, may not be fully appropriate to characterize the strength of MSW. However, at this time it appears that most MSW can be characterized in a soil-like manner as exhibiting "cohesion" as a result of the interaction of MSW particles (Mitchell and Mitchell, 1992) as well as frictional behavior. Thus, use of a cohesion "c" and friction angle "ϕ" to characterize strength is reasonable. Since undisturbed sampling and testing of MSW shear strength is very difficult, shear strengths used in design should be conservatively chosen from those reported in the literature and design should include calculations over a range of MSW strength values. Even then, good engineering practice would still probably require design of the shallow foundations using higher than conventional factors of safety to account for uncertainties in the MSW properties.

Deep Foundations. Driven piles are the type of deep foundation systems nearly always utilized to support larger structures constructed on closed landfills. The pile types commonly used at landfills are pre-cast prestressed concrete piles, steel H-piles, or steel pipe piles. Drilled piers or piles are rarely utilized because MSW is prone to caving, and casing must typically be used during drilling, thus increasing the construction cost. The drilling also produces large volumes of MSW from the subsurface which must be disposed at the site or at another landfill. Other landfills may be hesitant to accept MSW excavated from an old landfill owing to uncertainties regarding it's chemical composition and whether the excavated material may contain hazardous constituents at significant levels. In post-closure development of landfills the adage of *"Garbage in, garbage out"* does not seem to apply to excavated MSW. Additionally, excavated MSW can represent a health and safety issue for construction personnel and may also create an odor nuisance.

Design of pile foundations at landfills must include analysis of the following five main categories of issues:

* pile capacity, both vertical and lateral

* downdrag loads due to MSW settlement;

* constructability and construction impact on the landfill environment;

* corrosion resistance of pile ; and

* environmental protection and maintaining the integrity of MSW containment.

The last three items on this list are discussed in subsequent sections of this paper.

MSW rarely if ever provides sufficient strength and resistance to settlement to provide pile capacity, but instead generally develops negative skin friction or downdrag loads to piles. Vertical pile capacity is essentially always developed in soil or rock underlying the MSW mass at a landfill. Capacity can be achieved through to end bearing and/or skin friction in suitable soil or rock materials.

Design of vertical pile capacity is relatively conventional. First, a suitable geotechnical investigation should be completed to characterize the bearing materials underlying the landfill and their strengths. Conventional vertical pile capacity analyses can then be utilized to calculate pile capacity. Methods for these analyses are presented in many publications in the literature, for example in Prakesh and Sharma (1990) or Fellenius (1991).

Downdrag Considerations. A very significant issue that can greatly impact the overall design, and thus the type of pile foundations and installation procedures to be used at a landfill is the downdrag loads that the piles must carry. Downdrag, also known as negative skin friction, occurs when the settlement of the material surrounding a pile exceeds the downward movement of the pile shaft. As compared to the sizable settlement that may occur in MSW, downdrag may be fully mobilized by downward movement of surrounding materials with respect to a pile of only 15 mm (0.6 in.)(Vesic, 1977). This small amount of movement, as compared to the large settlements that can occur in MSW, readily leads to the conclusion that downdrag loads will usually be developed along essentially the full length of a pile through MSW unless some type of mitigation is provided.

Actual data on negative skin friction values for piles through MSW is very limited. However, the value of negative skin friction cannot exceed the shearing strength of the MSW itself. As was discussed previously, the shear strength of MSW is also quite variable. Fassett et al. (1994) present a good summary of MSW shear strength values which can provide an initial estimate of negative skin friction values. Rinne et al. (1994) report the results of a laboratory direct shear test between MSW and concrete tested for a specific site investigation which indicated an interface shear strength of 30 kPa (600 psf).

If the calculated downdrag loads are found to be appreciable, thus either requiring a significant increase in pile capacity or number of piles needed to support the loads, it is recommended that field pull-out tests be completed on a series of test piles to develop site specific shear strength values which can then be used to fine-tune the downdrag analyses. If possible, it is also desirable to instrument with strain gages the piles driven as part of the testing program to allow measurement over time of the actual downdrag loads which develop in the piles.

Downdrag loads can be large enough in extreme cases that nearly the entire capacity of a conventional pile may be required to carry the downdrag load. If the loads are significant, consideration must then be given to use of one or more mitigation techniques to reduce downdrag loads. These techniques, all of which may be applied through all or part thickness of the MSW layer, include:

- pre-drilling or spudding of the piles with a steel mandrel;

- pre-drilling or spudding of an oversize hole that is then backfilled with bentonite slurry;

- installation of an outer shell or casing, also known as a "double pile" system; and

- friction reducing coatings such as bitumen, installed through a cased or uncased hole developed by pre-drilling or spudding.

These techniques are shown schematically in Figure 3.

With the use of a pre-drilled or spudded hole, a steel mandrel is driven down through the MSW and then extracted. The hole is either left open or filled with bentonite slurry which can reduce the downdrag load significantly, provided that the hole is larger than the piles and does not collapse. However, installation of piles in this manner should not be used where pile lateral load capacity is necessary. Spudding may not be allowable if there is a concern regarding pushing small amounts of MSW ahead of each pile into the strata below the MSW. Pre-drilling also has the limitation of generating excavated MSW which must then be disposed of.

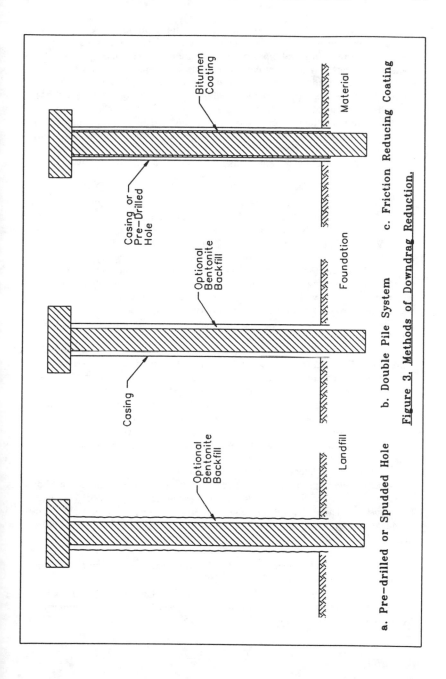

a. Pre-drilled or Spudded Hole   b. Double Pile System   c. Friction Reducing Coating

Figure 3. Methods of Downdrag Reduction.

The "double pile" method consists of an inner pile installed in an outer pile or casing which may be constructed using a variety of materials including steel pipe, corrugated metal pipe, or even plastic pipe. The inner pile is not in contact with the MSW and thus carries only the structural loads, while the outer pile carries the downdrag loads. The void between the inner and outer casing can be filled with bentonite for an increased level of protection, although this is generally not required. A limit of this system is the lack of lateral capacity, since the outer pile is not tied directly into the pile cap system so that it can settle in response to the downdrag loads. In theory, a spacer system between the inner and outer pile could be installed to increase lateral load capacity. However, it must be able to function without transferring significant downdrag loads to the inner pile.

Friction reducing coatings, such as a thick layer of bitumen, have a successful track record of reducing downdrag loads in non-landfill environments where they have been shown to reduce downdrag loads by up to 75 percent or more (Vesic, 1977). Load tests performed on bitumen coated pre-cast prestressed concrete piles driven through a landfill have shown lesser reductions on the order of 30 to 60 percent (Rinne et al., 1994). In order for the bitumen to provide a reduction of downdrag loads it must stay viscous over the range of temperatures which may occur with the MSW, and must not be scrapped off during installation. Pre-drilling or use of a temporary casing can reduce the potential for scraping off the coating. The chemical stability of the bitumen and possible effects on the environment are other considerations.

Lateral Load Considerations. In most pile supported structures, the resistance of the pile-soil system and pile caps or grade beams provides lateral load capacity. However, only that portion of the foundation system which can be ensured to remain embedded, and thus in contact with the soil or MSW, can be counted upon to provide resistance to lateral loads. Mitigation techniques used to reduce downdrag loads may thus greatly reduce lateral load capacity. Additionally as a result of sizable surface settlements, soil contact areas of grade beams and pile caps may be significantly reduced over time with an associated reduction in passive resistance to lateral loads. In order to compensate for these net reductions in capacity, pile caps may be increased in size, grade beams added or short drilled piers added to the overall foundation system to provide sufficient lateral load capacity. These elements all require careful structural engineering to ensure adequate structural stiffness in the lateral load resisting system, particularly at tie-ins.

The dynamic properties of MSW and response of landfills during earthquakes has not been studied to a great degree, although in general the seismic response of landfills has been very good (Orr and Finch, 1990) even for seismic events as large as the Northridge Earthquake of Richter Magnitude 6.7 (Stewart et al., 1994, Matosović et al., 1995). Seed and Bonaparte (1992) suggested using 80 percent of the static strength as a first approximation of dynamic shear strengths of MSW. Kavazanjian et al. (1995) provide an overview of MSW properties applicable to landfill development. Dynamic response analyses should be performed for significant structures to aid in estimating lateral load requirements.

Construction Considerations

A fundamental consideration in construction of foundations at landfills, as well as all construction at landfill sites, is the importance of the development and implementation of proper health and safety monitoring and protection procedures during construction. Landfills can pose a number of varying hazards which can become problematic during excavation or other construction operations such as pile driving. A primary hazard is landfill gas which can be explosive, and can carry other contaminants as well. Leachate within the MSW can also pose an exposure problem in some cases. Health and safety hazards associated with construction at landfills can generally be readily monitored and mitigated with relatively conventional techniques. The key is the preparation and implementation of a health and safety plan prepared by qualified personnel.

Shallow Foundations. Construction of shallow foundations on landfill sites usually follows normal techniques of construction. In rare cases foundation excavations encounter MSW which then must be properly handled and disposed. Generally shallow footings should bear upon soil, so excavated MSW may need to be replaced by engineered fill.

Deep foundations. As has been discussed previously, it may be necessary to utilize pile construction techniques to reduce downdrag. Additionally, the heterogeneous nature of MSW may result in obstructions which could block piles during driving, potentially resulting in pile breakage. Pre-drilling or other procedures may be necessary to deal both with downdrag and to reduce potential for pile breakage. Utilization of these techniques requires special considerations to deal with construction impacts. Pre-drilling or the use of drilled piers foundations may result in the production of sizable quantities of excavated MSW which must then be disposed in a proper manner. This waste disposal can be costly, and may also require permitting and testing which can delay construction. These impacts must be accounted for in construction scheduling.

Spudding, mandrel driving, or conventional driving of displacement piles or casing may result in pushing refuse ahead of the pile tips and out of the MSW layer. This may not be permitted by regulatory authorities, thus necessitating specific procedures to be employed to reduce or eliminate the potential for pushing MSW.

Jetting or the introduction of large quantities of liquid also may not be allowed by regulatory agencies owing to the possible impact on leachate quantities. Use of bentonite slurry can also be a problem if it must be disposed of after completion of pile installation, due to the potential that the slurry may have been contaminated.

In summary, construction operations for installation of deep foundations should be carefully planned to evaluate the possible impacts of pile installation on the environment of the landfill. Conventional techniques may not be allowed or may have to be modified to reduce impacts.

Corrosion Issues

Decaying MSW typically results in an environment which is corrosive to most types of deep foundations. Steel or concrete piles can be readily impacted by acids, chlorides, and sulfates which typically occur in MSW as it decays. Steel is more heavily impacted by chlorides, while acids and sulfates are more a concern for concrete products. Rinne et al. (1994) list the following methods of reducing corrosion potential for deep foundations. With concrete piles, methods include:

- use of Type II or V cement;

- maintain the lowest level of chloride in the concrete mix by limiting ingredients in the mix;

- increasing thickness of concrete coverage over steel reinforcement;

- adjusting maximum water to cement ratio to increase the strength of the mix;

- design the concrete mix for low water permeability and high density;

- epoxy coat steel reinforcement; and

- case the piles through the MSW and fill the void with concrete or bentonite slurry.

Methods to protect steel piles include the following alternative measures:

- thicken structural sections;

- utilize a protective coating, such as epoxy or polyurethane;

- concrete encasement of the portion of the pile in the corrosive environment;

• special corrosion resistant steel;

• cathodic protection; and

• installation of a separate casing.

Typically a corrosion specialist should be consulted during the project design stage to provide specific recommendations applicable to a specific set of site conditions.

Environmental Considerations

In the last few years, regulatory agencies have increasingly raised concerns on how the development of closed landfills, and in particular the installation of deep foundations, may effect the integrity of MSW containment at a landfill. Specific concerns have focussed on landfill liners or natural containment layers and the landfill cover. Two key concerns must be considered as part of the use of pile foundations. First related to water quality, a basic policy has emerged that the installation of pile foundations can not increase the potential for migration of contaminants from the MSW to ground water underlying a landfill. Second, related to air quality, is the policy that pile usage not lead to increased leakage of landfill gas from a site. The key issues are that piles not cause:

• increased infiltration of surface water into the MSW mass which could then increase leachate quantities;

• increased leakage of landfill gas out of the landfill; and

• leachate to migrate out of the landfill at a greater rate than would occur otherwise.

With regard to the base liner, either natural or constructed, as part of project design process sufficient information must be developed to demonstrate that use of pile foundations will not result in impact on ground water underlying a landfill. This requirement is typically satisfied by thorough investigation of the subsurface below the landfill to ascertain the existence of sealing layers that would be adversely impacted by installing piles through them. Basically, the piles must be shown to not act as a conduit for leachate out of the landfill. It would be difficult, if not impossible, to seal piles which were driven through a geomembrane or even a thin clay liner, and in some areas, including California, this is expressly limited by government regulations. Natural sealing layers require different considerations.

If a site is underlain by soft clays, data from routine soils tests such as grain size distribution, plasticity indices, consolidation, hydraulic conductivity, and shear strength can be used to characterize the scaling layer material properties. This data can then be presented along with a description of the pile installation process and the remolding and reconsolidation that the soft clay will undergo during and after pile driving to demonstrate to regulatory agencies that a seal will be maintained around the pile during and after driving through the sealing layer.

At a landfill where the natural sealing layer is stiff clay or fine-grained rock, other means have to be employed to ensure that a seal is developed around pile foundations at depth. Utilization of casings through the MSW which are then seated in the sealing layer, thus separating the piles from contact with the MSW may be necessary. It may be appropriate to fill the casings with bentonite to enhance the seal around the piles. These measures are costly and may limit the economic viability of certain structures at a landfill site.

Alternatively, there are also sites where no natural or constructed liner exists below the MSW. Usually in these cases, there is a high likelihood that the landfill does not contain appreciable quantities of leachate. Instead the leachate probably simply leaked out the base of the landfill. In this type of landfill, restrictions on pile usage may be lessened as long as it can shown that the piles will not enhance the movement of leachate out of the landfill. If ground water underlies a site without a sealing strata, installation of the piles to a limited depth below the landfill probably will not result in an adverse impact on ground water if the depth is well below the landfill. In this case it is a matter of demonstrating that the use of driven piles will not promote the movement of leachate or contaminants downward into ground water.

In a similar manner the design must provide for and demonstrate that foundations and structures that suitably tied into low permeability barrier layers that are incorporated into landfill covers to control escape of landfill gas and infiltration of precipitation. This is typically accomplished by careful detailing of the shallow foundations or pile caps and the installation of a geomembrane above the pile caps, below the floor slab, and tying in the barrier layer surrounding the structure.

Conclusions

The development of landfills with structures usually require carefully engineered shallow foundations or more commonly deep foundations. In order to successfully develop foundations on landfills, the design process must include the following:

•   a high level of detailed site evaluation;

- conservative design for ranges of MSW properties;

- inclusion of redundant project design features;

- careful construction with a higher than normal level of construction quality assurance; and

- careful planning and implementation of inspection and maintenance procedures and preparation of contingency plans for response to less than anticipated performance.

The design team and owner must work carefully together and the process is typically one of design evolution where different ideas are looked at in a variety of different ways, and the appropriate solution is selected. This design process cannot be hurried and the costs are typically higher than for conventional projects.

As part of the design, the design team must carefully consider the environmental impacts of the foundations on both air and water quality, and regulatory agencies must be shown that the foundations, and the project as a whole, will not result in significant environmental impact. Provided the owners, design team, and contractors are aware of the environmental constraints and objectives from the outset and work together with the appropriate regulatory agencies throughout the planning, design, and construction process, successful post closure development projects and possibly even enhanced environmental protection can be achieved. Failure to carefully plan design,and construct a project and then to maintain it after construction can result in severe aesthetic, environmental, and economic losses.

Appendix - References

Bjarngard, A. and Edgers, L., (1990) "Settlements of Municipal Solid Waste Landfills" *Proc. of the Thirteenth Annual Madison Waste Conference*, Madison, WI.

Dunn, R.J., (1993) "Successful Development of Closed Landfill Sites" *Proc. of Green '93 Waste Disposal By Landfill*, Sarsby ed. Balkema Rotterdam pgs. 527-534.

Edil, T.B., Ranquette, V.J., and Wuellner, W.W., (1990) "Settlement of Municipal Refuse," *Geotechnics of Waste Fills*, ASTM STP 1070, Landva/Knowles eds. pgs. 225-239.

Fassett, J.B., Leonards, G.A., and Repetto, P.C., (1994) "Geotechnical Properties of Municipal Solid Wastes and Their Use in Landfill Design," *Proc. Waste Tech '94*, Charlestown, SC, National Solid Waste Management Association

Fellenius, B.H. (1991) "Pile Foundations," *Foundation Engineering Handbook* 2nd Edition, Fang ed. van Nostrand Reinhold New York, NY, pgs. 511-536.

Howland, J.D., and Landva, A.O., (1992) "Stability Analysis of a Municipal Solid Waste Landfill," *Proc. Stability and Performance of Slopes and Embankments - II*, Vol. 2, Geotechnical Special Publication No. 31, ASCE, New York, NY.

Jessburger, H.L. and Kockel, R., (1991) "Mechanical Properties of Waste Materials," *XV Ciclo di Conferenze de Geotechnica di Torino*, Torino, Italy.

Kavazanjian, E., Matasovic, N., Bonaparte, R., and Schmertmann, G.R., (1995) "Evaluation of MSW Properties for Seismic Analysis," *Proc. of Geoenvironment 2000*, New Orleans, LA, ASCE, pgs 1126-1141.

Matasovic, N., Kavazanjian, E., Augello, A.J., Bray, J. D., and Seed, R.B., (1995) "Solid Waste Landfill Damage Caused by 17 January 1994 Northridge Earthquake," *The Northridge, California Earthquake of 17 January 1994*, Woods and Seiple eds., California Department of Conservation, Division of Mines and Geology special Publication 116, pgs. 61-69.

Mitchell, R.A. and Mitchell, J.K., (1992) "Stability Evaluation of Waste Landfills," *Proc. Stability and Performance of Slopes and Embankments - II*, Geotechnical Special Publication No. 31, ASCE, New York, NY.

Orr, W.R., and Finch, M.O., "Solid Waste Landfill Performance During the Loma Prieta Earthquake," *Geotechnics of Waste Fills*, ASTM STP 1070, Landva/Knowles eds. pgs. 22-30.

Oweis, I.S. and Khera, R.P., (1990) *Geotechnology of Waste Management*. Butterworths.

Prakesh, S. and Sharma, H.D., (1990) *Pile Foundations in Engineering Practice*, John Wiley & Sons, Inc., New York, NY, 734 pgs.

Rinne, E.E., Dunn, R.J., and Majchrzak, M., (1994) "Design and Construction Considerations for Piles in Landfills," *Proc. DFI 94 Fifth International Conference and Exhibition on Piling and Deep Foundations*. Bruges, Belgium, Deep Foundations Institute, pgs.2.2.1-2.2.5.

Seed, R.B. and Bonaparte, R., (1992) "Seismic Analysis and Design of Lined Waste Fills: Current Practice," *Proc. Stability and Performance of Slopes and Embankments - II*, Vol. 2, Geotechnical Special Publication No. 31, ASCE, New York, NY.

Sharma, H.D., and Lewis, S.P., (1994) *Waste Containment Systems, Waste Stabilization, and Landfills*. John Wiley & Sons, Inc. New York, NY 588 pgs.

Sheurs, R.E. and Khera, R.P., (1980). "Stabilization of a Sanitary Landfill to Support A Highway." *National Academy of Science, Transportation Research Record* 754: 46-53

Sowers, G.F., (1968) "Foundation Problems In Sanitary Landfills," *Journal Sanitary Engineering Division* ASCE Vol. 94 No SA 1.

Sowers, G.F., (1973) "Settlement of Waste Disposal Fills," *Proc. of the Eighth International Conference on Soil Mechanics and Foundations Engineering*, Moscow, Vol. 2, pgs. 207-210.

Stewart, J.P., Bray, J. D., Seed, R.B., and Sitar, N., (1994) "Preliminary Report on Principle Geotechnical Aspects of January 17, 1994 Northridge Earthquake," Report UCB/EERC-94/08, Earthquake Engineering Research Center, Berkeley, CA, 238 pgs.

Stulgis, R.P., Soydemir, C., and Telgener, R.J., (1995) "Predicting Landfill Settlement," *Proc. of Geoenvironment 2000*, New Orleans, LA, ASCE, pgs 980-994.

Tchobanoglous, G., Thiesen, H., and Vigil, S.A., (1993). *Integrated Solid Waste Management*. McGraw Hill.

Vesic, A.S., (1977) *Design of Pile Foundations*, Transportation Research Board, National Research Council, Washington, DC.

DESIGN OF
CIVIL INFRASTRUCTURE OVER
LANDFILLS

Max A. Keech[1]

ABSTRACT

This paper presents an overview of the design process for Civil Infrastructure on Closed Landfills. The design process starts with a review of Geotechnical & Historical information required by the Site Civil Engineer. Next, a procedure is described for identifying soft and hard edge boundary conditions which will govern the design. Design of finish slopes, utility systems, building connections and documentation of post-settlement conditions is discussed in detail. This paper concludes by defining Inspection and Maintenance requirements after construction.

INTRODUCTION

Historically, post-closure reuse of landfill sites has been limited to golf courses and parks or temporary/low value facilities. These uses are either relatively accommodating to ground settlements or have short design lives where the effects of settlement are often ignored.

Over the past 10 years, closed landfills in major urban areas have been increasingly viewed as offering potential for traditional urban developments, such as office parks and retail centers, due to the scarcity (and hence high value) of developable land. When developed correctly a win-win situation is achieved wherein a nuisance site is returned to productive use and the landowner has a vested interest in maintaining the property over time. Development pressure on more environmentally productive lands is also reduced through the "recycling" of closed landfill sites.

---

[1]Vice President/Principal, Brian Kangas Foulk,
540 Price Avenue, Redwood City, CA 94063

The past performance of developments on landfill sites has been somewhat spotty. Where the effects of settlement have been ignored or not adequately considered in the design process, serious cracking, utility failures and access difficulties have occurred. These failures can also threaten the integrity of the landfill cover by creating a pathway for liquid migration into or out of the landfill mass. Where the effects of settlement have been adequately addressed, successful developments have occurred which greatly reduce the potential for liquid migration, provide on-going maintenance and return the property to the community's tax rolls.

Traditionally, civil engineers have designed site infrastructure (loosely defined as everything outside the building envelope) as static, three-dimensional facilities. The design of site infrastructure on closed landfill sites requires a dynamic treatment which addresses the infrastructure's vertical movement over time due to the effects of settlement. While much information has been published on predicting landfill performance over time, virtually nothing is available on the design of site infrastructure to accommodate this performance. This text is intended to provide the Site Civil Engineer with a general overview of the dynamic design process and its inherent issues and potential solutions. It also provides the Geotechnical Engineer with a better understanding of how the Site Civil Engineer utilizes the geotechnical analysis, allowing a more focused and detailed treatment within the geotechnical report. This overview is based on recent design experience with six Northern California landfills and historic design practices utilized for development over highly compressible bay muds.

OVERVIEW OF THE DESIGN PROCESS

The Geotechnical Engineer is responsible for predicting the performance of a landfill site over time. Based on site specific borings and laboratory tests, the Geotechnical Engineer uses theoretical and empirical methods to predict site variables which affect the design of civil infrastructure. These include total anticipated settlement, location and horizontal extent of differential settlement, gas production, and other important aspects of the performance of the landfill site over time. Based on this analysis, the role of the Civil Engineer is to design the site improvements to accommodate the predicted performance of the landfill site. To do so the Civil Engineer must have a basic understanding of the variables that cause differing levels of settlement and other characteristics unique to landfill development.

When designing civil infrastructure for post-closure landfill development, a four-step design process should be employed. The four steps are:

I.   Analyze historical documents and geotechnical reports.

II.   Define site characteristics (boundary conditions).

III.   Design for predicted rates of differential settlement.

IV.   Define future inspection and maintenance requirements.

In preparing the design of civil infrastructure for post-closure use of a landfill site, three key design goals which must be achieved are:

1.   Achieve acceptable performance as the site settles. The design of all facilities must take into account the dynamic nature of the landfill site. Building entrances, utility connections, surface and utility slopes must be designed such that they are acceptable over the life of the development while settlement is occurring. Surface settlements may occur due to a combination of consolidation and decomposition of the landfill mass as well as consolidation of any compressible soils below the landfill mass. Infrastructure design must address the combined effects of total surface settlements as predicted by the Geotechnical Engineer.

2.   Avoid penetration of the low permeability landfill cap (barrier layer). Finished grades should be established such that subterranean improvements, such as footings and utilities are above the elevation of the barrier layer to provide effective protection against water intrusion into the landfill mass, as well as gas or leachate migration out.

3.   Minimize future maintenance requirements. By thoroughly analyzing the performance and settlement characteristics of the landfill site over time, the design can anticipate future performance and minimize future maintenance requirements. Again, the key is to recognize and plan for the dynamic condition of the landfill site.

The design methodology which follows is applicable to a closed landfill which is encapsulated with a low permeability layer (e.g. compacted or natural clay or geosynthetic clay liner). While a pile supported building is used to illustrate the range of design issues involved, a floating foundation would involve a similar, although less complicated design process relative to the site/building interface. The following design methodology addresses site improvements normally falling under site civil responsibility such as surface slopes and utilities, together with design issues at the site/building interface. Other important design considerations related to structural and corrosion issues, landfill gas collection and other facilities normally outside the responsibility of the site civil designer must also be addressed.

I. ANALYZE HISTORIC DOCUMENTS AND GEOTECHNICAL
    REPORTS

The first step in the design process is the analysis of historical documents to guide both the Geotechnical and Civil Engineers in understanding the design characteristics of a particular landfill site. Critical historical documents which should be analyzed include:

•   Original Landfill Grading Plan: An attempt should be made to obtain the original landfill grading plan. This document is often times the only reliable way of establishing landfill bottom elevations. This is critical in landfills which were originally constructed by excavation of a pit or open chamber since rapid changes in the landfill bottom elevation (typically at the perimeter of the pit excavation) generally represent boundary conditions where the depth of refuse, and hence future settlement, changes rapidly. These areas are particularly difficult to locate through field investigations due to the limited number of borings that are usually performed during the geotechnical site reconnaissance.

•   Sequence of Landfill Operations: Because decomposition and hence settlement is a time dependent activity, the sequencing of landfill operations for landfills which operated over a long period of time is important. Landfills can be encountered which are filled in horizontal layers or filled in cells over long, disconnected periods of time. Similarly, the original sequence of landfill operations may indicate differences in the composition of landfill waste, and hence differences in decomposition and settlement characteristics in different areas of the landfill.   While sequencing information can occasionally be obtained from written logs and operation reports, it is often available only through interviews with landfill operation personnel.

•   Existing Topography/Utilities and Landfill Gas Collection Systems: Existing topographic surveys with subsurface infrastructure should either be obtained or prepared prior to geotechnical investigations. This will allow the accurate establishment of bench elevations at the location of borings so that the barrier layer and refuse masses logged in the borings can be accurately defined by vertical elevation. Likewise, subsurface utilities such as gas collection systems, where present, will either need to be relocated or incorporated into the final design.

•   Previous Uses and Topography: It is quite common for landfill sites to serve as reclamation and storage sites after closure and prior to final post-closure development. These and similar past uses can have the effect of surcharging portions of the landfill site and altering anticipated settlement patterns. Sources of information on previous uses include historic aerial photos, topographic maps, city use permits, and most importantly, testimony of landfill owners and operators.

After obtaining and analyzing historical documents, the Site Civil Engineer should begin the process of collecting empirical data on the settlement characteristics of the landfill site. This use of the "Observational Method" is accomplished by setting settlement monitoring points (generally iron pipes), recording the initial elevation of the settlement monitoring points and re-recording over time. Numerous authors (Holm, 1993; Terzaghi and Peck, 1967; Wallace, 1995) have described general applicability and limitations of the Observational Method; however, very little specific information is provided as it relates to settlement monitoring. The Civil Engineer should consult with the Geotechnical Engineer to determine the number and location of the settlement monitoring points. In general, settlement monitoring points should have a higher density in boundary areas where settlement characteristics are anticipated to change rapidly (i.e. locations of large differential settlement). This includes boundaries of the landfill site where refuse depths change rapidly and areas which have undergone previous surcharging. For locations where settlement is anticipated to be relatively uniform, a grid spacing of approximately 45-60 meters (150 to 200 ft.) should be adequate. This leads to a monitoring point density of 5 to 10 points per hectare (1-2 pts/acre) plus the additional monitoring points defining boundary conditions. After initially establishing the monitoring point's elevation, intermittent readings should be recorded every two to four months for a period of no less than one and preferably two years to provide an adequate amount of empirical data to make predictions on future rates of settlement. Sowers (1973) reported that continuing landfill settlement is analogous to secondary compression of soil and was described as:

$$\Delta H = \frac{\alpha}{1 + e} H \log \left(\frac{t_2}{t_1}\right)$$

in which ΔH = total settlement, H= fill depth, e = void ratio, t = time, and α is a coefficient analogous to the compression index C_c.

Empirical data allows establishment of the combined coefficient ($\alpha/1+e$) based on site specific data. The Geotechnical Engineer can then utilize this empirical data with Sowers' formula to predict settlement characteristics of the site over time. Yen and Scanlon (1975) found correlation with this approach for landfill depths up to 31 meters (100 feet) based on field observations of three Los Angeles County landfills. Where empirical data is of sufficient density and duration to allow creation of a contour map of yearly settlement rates, straight line projections can provide an approximation of total anticipated settlement. Both methods assume a landfill site which will not undergo significant future environmental change (i.e. dry to wet, etc.).

Empirical monitoring coupled with a 20-year straight line projection provided the basis for the anticipated settlement contour map of a landfill in Colma, California shown in Figure 1.

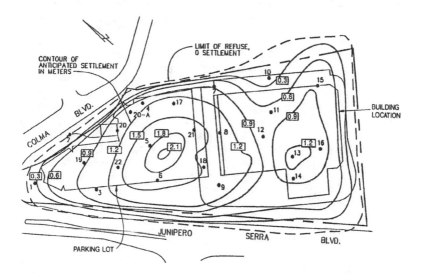

Figure 1.   Contour Map of Anticipated Settlement based
on Empirical Data

Ideally, settlement monitoring would begin with landfill closure so that the maximum amount of empirical data is available for the critical design process of predicting future settlements. This is rarely done, however, because the post-closure land use is not normally envisioned nor the design team in place at the time of closure. Closure regulations should be updated to require settlement monitoring after closure and eliminate this problem.

There are many components of the geotechnical report and analysis that provide predictions of site performance. Those that are most useful to the Civil Engineer are:

- Contour Map of Depth to the Barrier Layer and to Refuse: An accurate contour map showing the depth to the barrier layer and to refuse should be presented in the geotechnical report. This will allow the Civil Engineer to ensure that subterranean features such as foundations, footing, and utilities do not penetrate the barrier layer or lead to refuse excavation.

- Contour Map of Anticipated Settlement: A contour map should be prepared showing anticipated settlement, normally in 0.3 meter (1 ft.) intervals, for the landfill site based on existing surface grades. This contour map needs to account for the various boundary conditions contained within the site. Geotechnical Engineers may initially resist preparing a contour map of anticipated settlement due to the uncertainty in these predictions and potential liability associated with them. It is, however, the key element in designing site infrastructure to accommodate future settlements. For this reason, liability issues must be addressed early with the client. The contour map should be prepared based on the best available information including empirical settlement data and be accompanied by a sensitivity analysis which addresses the variability of key assumptions (e.g. - no change in environmental condition of the landfill mass). The contour map may require updating prior to the start of construction when additional settlement monitoring data is available.

- Additional Settlement per Foot of Additional Fill: It is often necessary to place additional fill on the landfill site to allow for placement of subsurface features without penetrating the barrier layer. The weight of this additional fill can cause additional consolidation (settlement) in the refuse and perhaps in the subsurface soils below the refuse. The geotechnical report should contain predictions of additional settlement per meter (or foot) of additional fill with appropriate language as to the variability of these predictions.

- General Analysis of Surcharge and Dynamic Compaction Options: A general discussion of surcharge and dynamic compaction options and their feasibility for a particular landfill site should be adequate for the initial geotechnical report. Should either of these options be deemed desirable during site specific design, they would then require further investigation and specific recommendations.

II.   IDENTIFICATION   OF   SITE   CONSTRAINTS   (BOUNDARY
CONDITIONS)

After completion of the analytical process, applicable site constraints must be identified. The site constraints that are particularly unique to landfill sites are commonly referred to as boundary conditions. If the entire landfill site settled at a uniform rate, the design of civil infrastructure on landfill sites would be no different from non-settling sites since the relative elevation or relationship between various improvements would remain constant after the effects of uniform settlement. However, differential settlement is always found on landfill sites due to the variable nature and depth of the refuse mass as well as the varying depths of fill required over the barrier layer. Where this differential settlement is relatively large or over a short horizontal distance, we commonly refer to this condition as a boundary condition. Boundary conditions occur due to natural changes in site variables, such as refuse depths leading to gradual ("soft edge") conditions and due to abrupt changes in settlement normally related to artificial boundaries created by deep foundations (i.e. pile-supported structures). These are referred to as "hard edge" boundary conditions.

A.   "Soft Edge" Boundary Conditions (Natural Boundaries)

"Soft edge" boundary conditions are areas of significant differential settlement occurring gradually over an identifiable horizontal distance. Typical causes of these soft edge boundary conditions which need to be identified are:

1.   Depth of Refuse/Topography of Landfill Bottom: Changes in the depth of refuse, oftentimes associated with changes in the topography of the landfill bottom, are one of the primary causes of "soft edge" boundary conditions. These are always present on the perimeter of the landfill site and may be found in interior locations for landfills that were phased or filled in a celled arrangement.

2.   Composition of Landfill Refuse: Significant changes in the composition of landfill refuse can lead to variations in both the rate of decomposition and consolidation and hence, variations in total settlement leading to "soft edge" boundary conditions.

3.   Age of Landfill Refuse: Where the age of landfill refuse changes significantly over the horizontal extent of landfill site, changes in the rate of decomposition can occur leading to "soft edge" boundary conditions.

4. Thickness of Landfill Cover: Variations in the thickness of landfill cover can lead to changes in the loading of the landfill mass and create "soft edge" boundary conditions. Variations in the thickness of landfill cover may pre-exist or may be created due to the final grading plan.

5. Previous Uses: As previously mentioned, landfill sites are often used for the recycling, storage and stockpiling of construction materials after their closure and prior to development activity. Large stockpiles placed over a long period of time can have a surcharge effect on the landfill leading to a lower rate of future settlement under the area where stockpiling previously took place.

Figure 2 schematically illustrates typical examples of "soft edge" boundary conditions and their effect on future finished elevations after settlement.

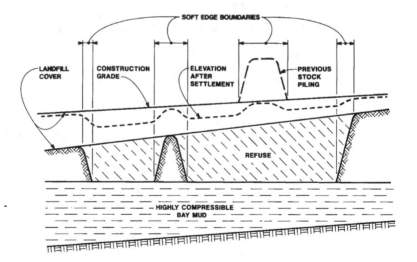

Figure 2. "Soft Edge" Boundary Conditions

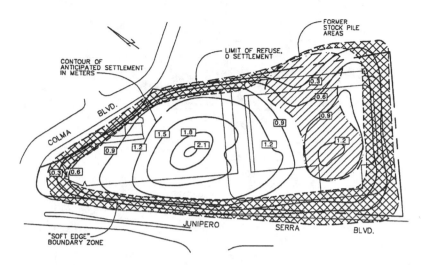

Figure 3. Major "Soft Edge" Boundaries at the Colma Landfill

Major "soft edge" boundaries are overlaid on the Colma settlement contour map in Figure 3. By reviewing the causes of differential settlement and hence "soft edge" boundary conditions listed above, it will be evident that many landfill variables are very difficult to define individually and are hence, best estimated collectively with the observational method. A well thought out settlement monitoring plan that produces a sufficient density of settlement data over a reasonable period of time will provide site-specific evidence of these combined effects on a particular site. In cases where the environmental condition of the landfill mass is likely to remain constant (i.e. dry or wet, etc.) and empirical data is obtained for 1 to 2 years or more, the reliability of predictions of landfill performance should be excellent.

B.   "Hard Edge" Boundary Conditions (Artificial Boundaries)

A special condition occurs at the edges of structures supported by deep foundations where an abrupt change in settlement occurs at the site/structure interface. This location of vertical shear is generally known as a "hard edge" boundary. This "hard edge" boundary is most commonly associated with the interface between pile-supported structures and site improvements supported directly by the landfill mass. Causes of vertical shear (Figure 4) which need to be identified in the second step of the design process include:

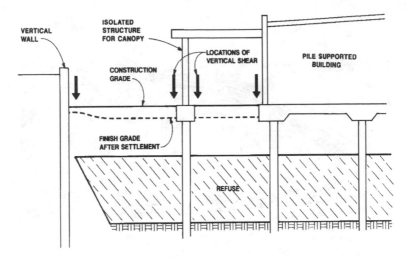

Figure 4. "Hard Edge" Boundary Conditions

1.   Edges of Supported Structures: Generally these will be pile
      supported structures or other structures with foundations
      extending through the landfill mass or supported by cantilever
      elements from beyond the landfill mass. Ingress/egress and
      utility connections crossing this boundary are of particular
      concern.

2.   Isolated Pile Caps and Grade Beams: Isolated pile caps and
      grade beams can occur for breezeways, signage, electroliers,
      and other isolated structural elements which are supported on
      deep foundations. These locations of vertical shear can be
      easily missed without adequate review of the structural
      drawings by the Site Civil Engineer. These isolated elements
      will tend to "rise" up from the site improvements as settlement
      occurs.

3.   Vertical Elements: Vertical site elements, such as retaining
      walls, may be required, generally near the perimeter of the
      site. Particular care should be given to dealing with the "hard
      edge" boundary conditions against these elements.

III.   DESIGNING FOR DIFFERENTIAL SETTLEMENT

After identification of site boundary conditions, we can begin designing for differential settlement. The three areas requiring design for differential settlement include finish slopes, site utilities, and pedestrian/utility connections to buildings.

A.   Finish Slope and Grading Design ("Soft Edge")

Conceptually we design for the ultimate finish slope desired after differential settlement. This is an integrative process achieved by first setting the anticipated construction slope required to achieve the desired ultimate slope. We then test the design by applying the anticipated settlement from the settlement contour map as well as the prediction of additional settlement for additional fill. Surface slopes must be designed to provide appropriate slopes after settlement. It is generally a good design practice to slope pavements and other surface improvements in the direction of increasing settlement to ensure no future reversals of surface flow directions. Changes of 1%-2% in future surface slopes due to settlement are not uncommon on deep landfills.

In general, roadways and other paving systems should utilize flexible materials such as asphaltic concrete and avoid or limit the use of Portland cement concrete and other non-flexible materials. Where concrete is utilized, adequate expansion and spacing joints should be provided to allow some flexibility for differential settlement. Subgrade considerations include an evaluation of the need for geotextile fabric or other materials below the base section to provide for bridging over localized settlements or potholing, which may occur within the refuse layer. In addition, consideration should be given to the prevention of water infiltration. Joints should be adequately sealed between differing materials such as asphalt and concrete curbs. Underdrains should be provided near the backs of curb and low points to pick up any excess water from irrigation or other sources. Finally, settlement monitoring points should be placed with the surface improvements to allow for future settlement monitoring.

Key design goals and solutions applicable to both soft and hard edge boundary conditions include:

•   Minimize Additional Fill: Finish grading design should utilize the minimum additional fill over the barrier layer required to avoid penetrations. Since the addition of fills will cause additional settlements in the refuse, and perhaps subsurface soils, minimization of additional fill will lower total overall and differential settlement.

- Avoid Penetration of the Barrier Layer: We must provide enough distance between finish grade and the barrier layer to allow for typical subsurface facilities such as foundations and footings, gas collection systems and normal utilities. Selective routing of utilities and other deep subsurface features is preferable to adding additional fill. For this reason, design of finish slopes and utilities must occur simultaneously.

- Avoid or Mitigate Difficult Boundary Areas: The Site Civil Engineer should consult with the architect and/or site planner to carefully locate the building, parking facilities and other hardscapes to avoid difficult boundary areas as an early consideration in planning the project.

- Mitigate Difficult Boundary Areas with Softscape (Landscaping): At soft (and hard) edge boundaries, where differential settlement is particularly severe, site planning which provides a generous amount of softscape or landscaping is particularly useful since disruptions from differential settlement will be less objectionable. It is important to limit subsurface irrigation lines in these areas because they are particularly vulnerable to cracking and leakage due to differential settlement. Flexible joints and/or placement of a drip or spray system above grade is a preferable design solution.

- Mitigate Difficult Boundary Areas with Surcharge: When time allows, surcharging is often useful to mitigate the effect of difficult boundary conditions where important facilities such as buildings and site entrances are located. The Geotechnical Engineer should provide the Site Civil Engineer with height/time/settlement relationships so that the most efficient surcharge system can be developed. The effect of surcharging on the barrier layer should be analyzed and consideration given to tapering the surcharge at its perimeter to minimize abrupt changes in future settlement.

- Dynamic Compaction: Occasionally dynamic compaction of refuse is an option for localized fingers of waste and certain other specific conditions. Dynamic compaction is rarely useful, however, within the main body of the landfill since it inevitably damages the barrier layer and creates its own boundary conditions at the limits of the dynamic compaction effort.

• Unloading of Refuse: Occasionally landfill sites or portion of sites are encountered which have a large amount of cover over the barrier layer. It may be viable to remove some of this cover, in effect unloading the refuse layer by removing an amount of earth equivalent in weight to the dead load of the replacement structure and other facilities. This should reduce or eliminate the effects of additional settlement due to additional structural weight. Care should be taken, however, in assuming that existing time rates of settlement would be further reduced by unloading since very little published data is available on which to estimate any possible reductions associated with unloading. A post-unloading settlement monitoring program over an extended period of time would be required to accurately predict the effects of unloading.

• Refuse Relocation: Where regulatory agencies will allow refuse relocation, it can be a particularly useful technique in areas where large differential settlements are encountered due to changes in refuse depths. This technique was particularly successful on a Northern California site where a baseball field had been built over one section of a landfill in Belmont after the placement of only five feet of refuse. Refuse placement continued on the remainder of the site reaching a depth of approximately 15 feet with one area reaching 22 feet in depth. During redevelopment of the site more than a decade later, the high refuse area was lowered and the refuse used to fill in the low lying baseball field area, creating a relative balance in the overall depth of refuse and future settlements.

B.   Site Utility Design

Site utilities should be designed for ultimate slope caused by anticipated settlement from the settlement contour map as well as for additional settlement from additional fill. Particular care must be given to gravity utilities so that reversal of flow direction is not encountered as settlement occurs. Designing for gravity utilities which must exit the site to municipal facilities inevitably must contend with the bowl phenomenon (Figure 5).

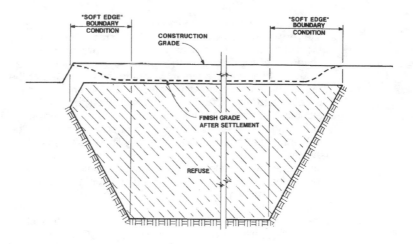

Figure 5. The "Bowl" Phenomenon

"Soft edge" boundary conditions exist around the perimeter of the landfill where the refuse depth changes abruptly and settlement becomes negligible outside the landfill site. Gravity utilities must be designed with adequate slope at construction so that after settlement, reversal does not occur across these soft edge boundary conditions. In general, design slopes are increased to accommodate slope reductions due to differential settlement and provide a "factor of safety" if the predicted differential settlements are not conservative enough to account for this. Appropriate factors of safety must be carefully chosen based on experience and will generally decrease with increased settlement monitoring periods.

Other utility design considerations which are illustrated in Figure 6 include:

- Minimize Utilities within the Landfill Area: Often this can be accomplished by combining utilities in a common utility chase. Sheet flowing of surface waters where possible to minimize the total amount of storm drainage piping is also advisable. Flexible piping materials such as PVC and polypropylene should be utilized to provide for anticipated differential settlement. Rubber gasketed joints rather than glued or solid joints should be utilized to allow additional flexibility in utility systems.

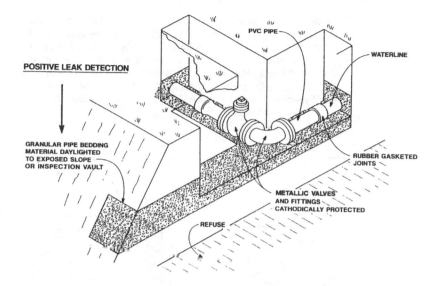

Figure 6. Utility Site Design

- Positive Overflow and Leak Detection: Providing a positive overflow and leak detection system should be an inherent part of the design for all pressure utility systems. For utilities subject to large differential settlements over relatively short distances, double encasement in a PVC carrier pipe is recommended.

- Above Grade Utility Installations: In some instances, utilities such as irrigation lines can be effectively placed above grade with little harm to the aesthetic integrity of the site development. In addition, above grade utility installations are appropriate for difficult boundary conditions such as steeply sloped edges of landfill sites.

- Barrier Layer Penetrations and Gas Migration Barriers: While the utility designer should attempt to minimize and/or eliminate any penetrations of the barrier layer, sometimes penetrations are necessary or may be inadvertently encountered in the field during construction. This condition should be accommodated in a fashion similar to Figure 7, where an impermeable liner is used within the utility trench and effectively tied back into the barrier layer. Styrofoam or similar blocking should be used on the landfill side of this liner to avoid penetrations from objects contained within the refuse layer. Geosynthetic Clay Liners (GCL) should be utilized when punctures are of concern. Where the potential for gas migration into the utility trench exists, gas migration barriers such as plugs or collars of impermeable materials should be provided at regular intervals.

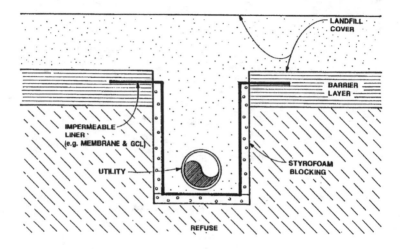

Figure 7. Utilities Penetrating Landfill Barrier

- Cathodic Protection: For many landfill developments cathodic protection of metallic fittings such as water valves may be necessary due to the potentially corrosive environment. The designer should consult with the Geotechnical or Corrosion engineer as to the needs for protection on each individual site. Cathodic protection through the use of buried anodes can be difficult to obtain due to the requirement of finding adequate burial depth for the anodes above the landfill cap.

C.   Building Connections ("Hard Edge")

Special consideration must be given to the design of connections to buildings supported on piles.   Where accessways and utilities enter or leave the building, vertical dislocation caused by vertical shear as site improvements settle at the building interface must be accounted for. At these locations it is not uncommon to encounter a vertical dislocation of 0.5 or more meters (1½+ ft.). For non-pile-supported buildings, many of the design solutions discussed below are applicable, but the condition is much less severe. Under non-pile-supported conditions, differential settlement is often reversed with the building (due to its dead load) settling more than the ground around it.

Methods used to mitigate boundary conditions at building interfaces include:

• Hinged slabs to accommodate vertical dislocations: Figure 8 schematically shows a hinged slab at a pile supported building entrance located on a landfill site. The slab is hinged at the building face to allow angular rotation. The hinged slab is generally designed for a beam loading condition with no center slab support.   Ultimate slope of the hinged slab must be designed to accommodate the desired access, (i.e. handicap, vehicle, user requirements). Based on anticipated settlement and maximum acceptable slope after settlement, the length of the hinged slab can be calculated.   On sites with large settlements, it may be necessary to initially have the hinged slab slope towards the building while providing appropriate drainage near the end of the hinged slab to avoid water intrusion into the building.   Note that a gallery or similar drainage system is required on both sides of the hinged slab since the drainage path will reverse itself after settlement occurs. The problem of vertical dislocation occurs in virtually all building access locations and, in general, a hinged slab is the appropriate design solution. Special design considerations must be given to hinged slab joints and sides where interface movements must be thoroughly analyzed.

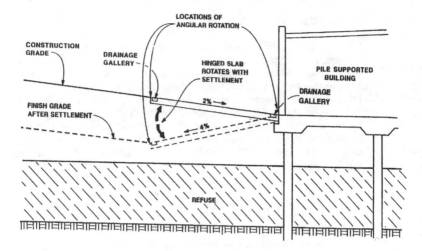

Figure 8. Hinged Slab at Building Entrance

- Vertical Dislocations at Footings, Grade Beams and Vaults:
Vertical dislocations should be anticipated along the entire
perimeter of the building. Where footings extend beyond the
face of the building, this dislocation may be particularly
noticeable, leading to exposed structural elements as
surrounding site improvements settle. Exposed "gaps" can
occur along the structural system leading to water infiltration
below the surface improvements. The exterior face of grade
beams and footings should always be formed (vs. poured neat)
to provide a smooth slip surface for the settling site
improvements. Where possible, grade beams exteriors should
be set with their exterior edge flush with the building face. It is
important that the depth of grade beams and other items be
sufficient to be fully covered after anticipated site settlement or
that skirts or other mechanisms be provided to avoid creating
an opening to voids under the building. A particularly difficult
problem is caused by isolated grade beams which may be
encountered at breezeways or other pile supported items such
as electroliers, etc. The site planner should try to avoid these
isolated structural systems since mitigation of vertical shear is
difficult and these features will tend to raise up from other site
improvements as settlement occurs.

- Vertical Dislocations at Utility Connections: Utilities that exit the building either above or below the structural slab must be designed to accommodate anticipated settlement. Since utilities within the building footprint are generally connected to the structural system, they will not settle, while the utility immediately adjoining the building will encounter the full anticipated site settlement. This vertical dislocation should be handled through a flexible utility connection contained within an inspection and leak detection vault. Examples of flexible utility connections for both pressure and gravity systems are shown in Figure 9. Pressure utility systems such as domestic water and fire lines must be suitably anchored on both the building and site side of the flexible utility connection so that pressure forces at the utility bends do not expand the flexible connection and cause pipe distress above or below the flexible connection. Gravity utility connections are somewhat easier to handle by overlapping the building utility line with the site utility. The length of this overlap, or extension potential of pressure connections, must be specified based upon anticipated settlement at the building's perimeter. Ease of inspection and access should be provided since these systems require ongoing observation and maintenance and may require replacement if settlement exceeds anticipated levels.

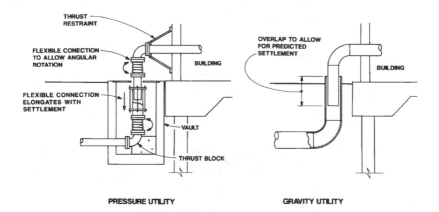

Figure 9. Flexible Utility Connections

- Design Considerations below the Structural Slab: Special design consideration must be given to utility systems carried below the structural slab. Typically these utilities are supported by hangers placed within the structural slab during its pour. During settlement, loading of the pipe hung from the slab can occur due to the backfill within the utility trench. Over time this can lead to hanger failure and the utility system becoming dislocated from the slab, leading to leaking where it rises to enter the building. To mitigate this problem non-cohesive backfills such as pea gravel should be utilized in utility trenches below the structural slab. Other successful systems include not backfilling the utility below the slab and placing plywood or other materials over the trench to support the structural pour. Hangers should be of a strong, but non-corrosive material and should have adequate redundancy so that an individual failure does not cause the utility to dislocate from the slab. A typical sub-slab utility design is shown in Figure 10. Unfortunately, the use of design/build specifications rather than a thorough design for sub-slab utilities has led to numerous instances of problems on both landfill and other land development sites built over highly compressible soils.

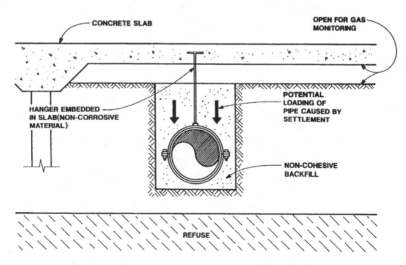

Figure 10. Utilities Suspended below Pile-Supported Slabs

D.   Post-Settlement Design Documentation

After preparation of the final design documents showing finish surface slopes, utility grades and building connections, the Site Civil Engineer should prepare post-settlement design documents showing the anticipated finish surface slopes, utility slopes, building connections, and other aspects of site improvements after anticipated settlement as predicted by the Geotechnical Engineer. This post-settlement design documentation is a critical element of landfill design and when done in sufficient detail will provide a "reasonable" check on the suitability of the design for its intended purpose. Uncertainty will still exist due to uncertainties in predicting settlement and the potential for unforeseen, localized conditions. Critical areas should have appropriate factors of safety and/or retrofit flexibility. Potential risks should be discussed early with the client and limits of liability established.

IV.   DEFINE INSPECTION AND MAINTENANCE REQUIREMENTS

While the design of site infrastructure should be performed to minimize future maintenance needs, inspection and maintenance is an ongoing requirement for any facility on a landfill site. Inspection and maintenance (I & M) must be part of the ongoing operation and maintenance (O & M) of the facility after its construction. To assist the O & M personnel after occupancy, it is imperative that the engineer prepare an I & M manual for their use. This manual should contain both inspection procedures and timing as well as notification procedures when other than routine maintenance is required. Items which should be addressed in the I & M manual include:

- Settlement Monitoring: Settlement should be monitored on a regular basis and compared to anticipated site settlements. Settlement monitors which were installed as part of the site improvements should be surveyed on a regular basis beginning with twice a year for the first two years and then once every two to three years following. The manual should provide a graph of anticipated settlement for selected monitoring points over time, allowing maintenance personnel to easily plot settlement against that anticipated. A notification procedure should be provided for settlements which exceed anticipated ranges and may lead to failure of flexible utility systems or unacceptable slopes at building entrances, etc.

- Pavement Condition: Pavement condition and surface slopes should be visually checked twice a year. Small cracks or voids in areas where water infiltration could occur should be sealed on a regular basis.

- Hinged Slabs: Hinged slabs should be visually inspected twice a year with special attention paid to the leading and tailing ends where angular rotation occurs.

- Utility Connections and Vaults: Utility connections should be examined where they connect to the building or run off-site. A visual inspection should be made to ensure that the flexible utility connection is moving adequately with settlement and that no distress is occurring to the pipe above or below this connection. Intermediate vaults should be checked for leaking and general condition of the piping system, where visible.

- Utility Televising: Storm and sewer lines should be televised once every five years to check for sagging or other obstructions which may occur within them.

SUMMARY

The successful design of site infrastructure for post-closure landfill developments can be accommodated through a collaboration of the experienced Civil and Geotechnical Engineer. In so doing we must:

- Predict anticipated settlement based on adequate observation.

- Recognize practical limitation in the geotechnical analysis and address liability issues.

- Recognize and identify all soft and hard edge boundary conditions.

- Design for ultimate slope and elevation based on anticipated differential settlement. Both construction and anticipated post-settlement slopes and elevations need to be defined and documented.

- Minimize facilities over difficult boundary conditions.

- Avoid barrier layer penetrations.

- Minimize future maintenance requirements and water infiltration.

- Maximize the ease of inspection/access to utility systems and provide positive leak detection.

- Define inspection and maintenance requirements.

While post-closure development of landfill sites with permanent facilities is a relatively new undertaking, there are numerous examples of both successful and unsuccessful designs. After reviewing numerous problems involving landfill sites with inadequate infrastructure designs, it is apparent that the majority of these problems occur in locations where differential settlements (hard and soft edge boundary conditions) were not identified and no design solutions were provided, rather than designs that did not adequately accommodate the actual vs. anticipated levels of settlement. For this reason it is apparent that attention to detail and adequate design energy are the most important components in a successful landfill design.

REFERENCES

Holm, L.A., "Strategies for Remediation", Geotechnical Practice for Waste Dispooal, Ed Daniel, D.E., Chapman & Hall, London, 1993, pp. 306-309.

Sowers, G.F., "Settlement of Waste Disposal Fills," Proceedings, 0th International Conference on Soil Mechanics and Foundation Engineering, Moscow, Union of Soviet Socialist Republic, 1973, pp. 207-210.

Terzaghi, K. and Peck, R., "Soil Mechanics in Engineering Practice", 2nd edition; Wiley, New York, N.Y., 1967, pp. 294-295, 632-643.

Wallace, W.A., "A New Framework for Hazardous Waste Remediation", Proceedings, ASCE Specialty Conference Geoenvironment 2000, New Orleans, Feb. 24-26, 1995, pp. 1630-1645.

Yen, B.C. and Scanlon, B., "Sanitary Landfill Settlement Rates", Journal of the Geotechnical Engineering Division, ASCE, Vol. 101, No. GT5, May 1975, pp. 475-487.

FOCUSED REMEDIAL INVESTIGATION/FEASIBILITY STUDIES FOR CERCLA-LISTED LANDFILL SITES

MATTHEW J. VELTRI[1]
Affiliate Member, ASCE

[1] Supervising Engineer
McLaren/Hart Environmental Engineering Corporation
8500 Brooktree Road, Suite 300
Wexford, Pennsylvania 15090

ABSTRACT

A protocol is presented for preparing a focused remedial investigations/feasibility study (RI/FS) for Comprehensive Environmental Response, Compensation and Liability Act (CERCLA) listed landfill sites. This document provides information pertaining to a broad-based understanding of the RI/FS process for National Priorities List (NPL) landfill sites. The adverse environmental concerns associated with landfills are common to many sites. Because of these similarities in landfill characteristics, similar remedial technologies ("presumptive remedies") may be considered at different sites. Presumptive remedies have been developed to expedite future remediation at these sites. Presumptive remedies are preferred Federal or State technologies for common categories of sites, based upon historical patterns of remedy selection and technology performance information.

The critical site characterization issues related to performing the remedial investigation at CERCLA listed landfills have been provided. Additionally, the recommended procedures necessary to develop and evaluate individual remedial technologies and remedial alternatives for landfill remediation have been summarized. The guidelines provided herein are based upon practical experience associated with preparing focused RI/FSs and information extracted from the applicable Environmental Protection Agency (EPA) and other regulatory guidance documents.

INTRODUCTION

Approximately 20 percent of the sites listed on the EPA-NPL are municipal landfill sites (OSWER Publication 9355.3-11, February, 1991). Many of these sites share similar site conditions and potential environmental concerns. Previously, engineers and scientists were required by the CERCLA "Superfund" process to evaluate a full spectrum of technologies for remediation of landfill sites. A focused RI/FS can be performed to assess the appropriate technologies and alternatives in order to streamline the evaluation process. Focusing the RI/FS can result in time savings and ultimately in a more cost effective process. General remediation guidelines are provided herein, however, specific sites may exhibit unique characteristics that require remedial action beyond the scope of this report.

The process of performing a RI/FS was initiated through the implementation of the National Contingency Plan (NCP) and the document titled *Guidance for Conducting Remedial Investigations and Feasibility Studies Under CERCLA* (OSWER Publication 9355.3-01, October, 1988). Streamlining the RI/FS and associated remedy selection for specific classes of sites with similar characteristics has been a focus of the U.S. EPA Office of Emergency and

Remedial Response (OERR) since 1993.

The EPA has initiated the Superfund Accelerated Cleanup Model (SACM) to make Superfund management more timely and efficient through early action and long-term action remediation (OSWER Publication 9203.1-051, December, 1992). Containment is considered to be an appropriate long-term (i.e. non-time critical) response to an identified release. Non-time critical actions can be initiated more than six months after the determination that a response action is necessary. The EPA has developed a framework for evaluating and selecting remedial technologies and alternatives through the preparation of a Engineering Evaluation/Cost Analysis (EE/CA) for non-time critical actions (OERR Publication 9360.0-32, August, 1993). An EE/CA is essentially a focused feasibility study. The Federal or State Remedial Project Manager (RPM) will make the decision concerning whether a specific site is appropriate for the non-time critical actions. Generally, this determination is supported by performing a streamline risk assessment at the site. In some cases, the RPM may elect to recommend the EE/CA process for sites that have not been listed on the NPL, but appear to be a potential environmental concern. The EE/CA process may be recommended by the RPM for categories of sites with presumptive remedies (described below).

Because of these similarities in landfill characteristics, similar remedial technologies ("presumptive remedies") may be considered at different sites. Based upon the information acquired from the remediation of similar sites, the EPA has developed presumptive remedies to expedite future remediation at these sites. Presumptive remedies are preferred Federal or State technologies for common categories of sites, based upon historical patterns of remedy selection and technology performance information. The presumptive remedy approach is included as part of the SACM directive. This SACM directive identifies containment as the primary on-site presumptive remedy for CERCLA-listed municipal landfill sites (OSWER Directive 9355.0-47FS, September, 1993) along with collection and/or treatment of the landfill gases. Additionally, source area groundwater control, leachate collection and treatment and institutional controls may be included as part of the presumptive remedy. Also, the EPA has identified containment technologies as the appropriate technology for wastes that pose a relatively long term threat or where treatment is impractical (OSWER Directive 9355.3-11FS, September, 1990). Recommended procedures for evaluating these presumptive remedy technologies along with evaluation criteria for addressing common environmental concerns identified by the author are provided herein.

The presumptive remedy does not address exposure pathways outside of the source area (landfill), nor does it typically include potential long term groundwater response actions. A response action for exposure pathways outside of the source area (if present) may be selected together with the presumptive remedy or as a separate operable unit. Operable units are discrete phases of remediation designated by the EPA. CERCLA-listed landfill RI/FSs are often separated into two operable units. The first operable unit typically addresses on-site remediation measures (e.g. containment). The second operable unit may address long-term impacted groundwater issues, if present.

REMEDIAL INVESTIGATION

The remedial investigation (RI) is an iterative process and several investigatory programs may be necessary to focus the site characterization efforts. Therefore, rescoping of the RI process may occur throughout the RI/FS process. Additionally, scoping of the activities required to obtain pre-design information may be identified following completion of the RI/FS as part of the Remedial Design Work Plan. Site characterization data is used to: assess the potential risks posed by the site; to gather the information required for remedy selection; and, subsequently to perform the remedial design activities. Typical remedial investigation procedures for CERCLA-listed landfill sites are provided below.

General Slte Conditions

Characterization of the overall site conditions is necessary to aid in determining the most appropriate remedial action. Characterizing the general landfill site conditions may include evaluating the information available relative to the following items:

- Site mapping and surveying data
- Types of waste material disposed in the landfill
- General landfill characteristics including, existing landfill covers, steep side slopes, erosion, slope failures, etc.
- Boundary of landfill contents
- Surrounding manmade surface features including homes, buildings, property lines, fencing, utilities, roadways, etc.
- Climatological data

General site information may be obtained through regulatory agency records or any other historical data that may be available. Site mapping is an essential component for initiating a RI and completing the Feasibility Study (FS). If adequate site mapping is not available, aerial mapping will be necessary along with any supplemental surveying to obtain baseline site conditions. Once adequate site mapping has been established, RI activities can be performed. Significant landfill and surrounding surface features should be recorded on the site map. If as-built data is not available for the existing containment system (if present), a field investigation may be necessary. This may include a geotechnical investigation of soil covers to assess the thickness and quality of the cover. If the limits of waste disposal are unknown, these limits must be established based upon review of historical aerial photographs, and field investigations (i.e. geophysical studies, soil borings and/or test pits).   Climate and precipitation data in conjunction with other site data can be utilized for containment system selection.

Isolated Areas of Concern ("Hot Spots")

Isolated areas of concerns should be examined if historical documentation or any physical or chemical data exists to support their presence. Hot spots may be identified within the confines of the landfill or may be separate areas outside of the landfill boundary. Hot spots may include: surface impoundments; liquid storage areas ("lagoons"); isolated disposal areas; and areas of drum or container disposal. Hot spot delineation may be conducted based upon reviewing available data, physical or chemical procedures, and based upon evaluating the existing site conditions. Hot spots may also include areas of impacted soils and sediments as described below.

Characterization of soils and sediment hot spots should be concentrated in areas of greatest concern.   This includes soils/sediments with obvious visual impact (e.g. staining from leachate seeps) or areas with historical significance (e.g. soils adjacent to known hot spot disposal areas or discharges).   Field screening techniques can be useful for preliminarily identifying areas of impacted media.   If existing information does not provide a basis for predicting sampling locations, a grid pattern may be selected for sampling.

Groundwater/Leachate

Characterization of the site geology and hydrogeology with respect to the location of the landfill contents is necessary to assess if remedial activities for groundwater/leachate will be required. Leachate generation may occur as a result of infiltration of precipitation, surface water infiltration and/or through groundwater recharge. Groundwater/leachate characterization includes gathering general site data, recommending the installation of monitoring wells, and/or theoretical evaluations.

The RI for groundwater/leachate also includes establishing surface water drainage trends, leachate characteristics (e.g. sampling and analyzing leachate seeps), and identification of local aquifer conditions. Depending upon the presence of monitoring wells, it may be appropriate to recommend up-gradient or down-gradient monitoring wells.   Theoretical evaluations may include groundwater modeling programs such as "Modflow" (McDonald, et. al., USGS, 1988) to determine the potential for the groundwater to contact the landfill contents.   Leachate modeling as a result of surface water infiltration may be performed by utilizing the Hydrologic Evaluation of Landfill Performance (HELP) Model (Schroeder, et. al, EPA/600/R-94/168a, September, 1994).

Landfill Gas

Landfill gas characterization may include both theoretical and field evaluation techniques. Theoretical evaluations include estimating the methane and volatile organic generation potential of the landfill contents (Federal Register, AD-FRL-3780-9, May 30, 1991). This potential can be estimated by gathering information such as the types of materials disposed in the landfill, the age and depth of the landfill contents.

Field investigations are performed to determine the potential for off-site migration of landfill gases and/or to determine the quality of landfill gases. This includes ambient air monitoring along with soil gas sampling around the perimeter of the landfill. Soil gas monitoring is typically performed with temporary gas monitoring probes in conjunction with field sampling and analysis. Landfill gas sampling and analysis can also be performed at the discharge points for existing gas management systems. Gas monitoring probes can also be utilized to evaluate the quality of the gas within the landfill (Federal Register, AD-FRL-3780-9, May 30, 1991).

Wetlands, Surface Water and Floodplains

Numerous landfills have been constructed above or adjacent to wetlands or surface water (streams, rivers ponds and lakes). Several issues may arise under these circumstances: constituents of concern adversely impacting the wetlands/surface water, loss of wetland habitat, and physically relocating surface water to accommodate construction.

The requirements of the NCP, as amended by the Clean Water Act, include the identification of the possible existence of jurisdictional wetlands associated with the site, and if present, delineation of the extent of the wetlands. The potential environmental impacts to wetlands and surface water should also be identified. Floodplains should also be identified through the review of drainage basin studies and flood insurance maps.

Baseline Risk Assessment

The purpose of the baseline risk assessment is to determine if the potential risks to human health and the environment warrant remedial action at a site (OSWER Directive 9355.3-11FS, September, 1990). The baseline risk assessment is also utilized by the regulatory agency to reinforce the finding of a potential imminent and substantial endangerment, if required as part of an enforcement action. This streamlined risk assessment is performed by the EPA to determine if early action of critical landfill concerns should be addressed.

REMEDIAL TECHNOLOGIES

This section provides the criteria necessary to develop remedial technologies as part of the feasibility study process for CERCLA-listed landfills. The remedial technologies are ultimately selected based upon remedial action objectives established for the site and the results of the RI. Additionally, remedial technology selection is based upon the applicable or relevant and appropriate requirements (ARARs) for a specific site. A description of remedial action objectives and ARARs are presented below, followed by a description of the typical remedial technologies for CERCLA-listed landfills. The technologies discussed represents a focused subset of the plethora of technologies available. Recommended procedures to focus the technology evaluation and selection process have also been provided.

Remedial Action Objectives

Remedial action objectives (also known as removal action objectives as part of the EE/CA process) are developed for a site to identify media-specific goals to protect human health and the environment. Specifically, these objectives are developed to reduce or eliminate the potential exposure pathways that the constituents of concern may have on human or environmental receptors.

ARARs

The Superfund Amendments and Reauthorization Act of 1986 (SARA) program contains a provision that remedial actions must at least attain Federal and State ARARs (ORD Publication EPA/625/4-91/025, May, 1991). Under Section 300.415(i) of the NCP, the selection of a remedial action at National Priorities List (NPL) sites must comply with ARARs of Federal and State environmental laws, to the extent practicable considering the urgency and scope of the action. These environmental laws include those established by EPA, and other federal agencies and those established by the State agencies.

ARARs are classified according to the NCP Section 300.5 as:

"Applicable Requirements" are the cleanup standards, standards of control, and other substantive environmental protection requirements, criteria, or limitations promulgated under Federal or State law that specifically address a hazardous substance, pollutant, contaminant, remedial action, location, or other circumstance at a CERCLA site.

"Relevant and Appropriate Requirements" are the cleanup standards, standards of control, and other substantive environmental protection requirements, criteria, or limitations promulgated under Federal or State law that, while not "applicable" to a hazardous substance, pollutant, contaminant, remedial action, location, or other circumstance at a CERCLA site, address problems or situations sufficiently similar to those encountered at the CERCLA site that their use is well suited to the particular site.

Considering the origin and objective of the requirement is recommended. For instance, while Resource Conservation and Recovery Act (RCRA) regulations may not always be applicable to closing undisturbed hazardous waste in place, the RCRA requirement for closure by capping may be deemed relevant and appropriate (refer to Landfill Containment Requirements).

ARARs can be placed into three categories: chemical-specific, location-specific and action-specific.

Landfill Containment Requirements

The major technologies that are typically considered at a minimum for containment include:

- RCRA Subtitle C Cap (composite barrier cap)
- RCRA Subtitle D Cap/State-required Cap (typically single barrier cap)
- Utilize or upgrade existing cover/cap (if present)
- Erosion and sediment control

Containment technologies are designed to provide a barrier to prevent direct contact with the landfill contents, minimize the percolation of storm water into the landfill contents, control landfill gas emissions, and control erosion. The potential for excavation and grading of landfill contents and surrounding topography should be considered and discussed. Grading of landfill contents may be required to achieve stable slope conditions for placement of the cap. Climatic conditions and topography can be used as input parameters for the HELP Model to determine if a stormwater drainage layer is necessary within the cap and to compare the general permeability performance of different capping systems. Climatic data can also be utilized to determine frost protection requirements (if any) for the cap.

Initially, the construction and integrity of the existing containment system (if present) should be analyzed to determine if the system meets the remedial action objectives and ARARs. Additionally, existing containment systems may be adequate to serve as a component (e.g. bedding layer) of the cap selected. Design engineers may also propose a hybrid of the regulatory agency recommended capping systems. This may include proposing innovative capping components such as textured geomembrane and synthetic drainage net. The preliminary stability characteristics of the proposed cap may be estimated by using the infinite slope stability evaluation (Dunn, et. al., 1980).

RCRA Subtitle C requirements are applicable for a Superfund remedial action if the following conditions are met (OSWER Publication 9355.0-49FS, September 1993):

- The landfill contents contain RCRA-characteristic waste and the waste was disposed after November 19, 1980, or;

- The response action constitutes disposal under RCRA (i.e. disposal back into the original landfill).

If RCRA Subtitle C is not applicable at a site, Federal or State Subtitle D regulations are typically considered relevant and appropriate for CERCLA response actions.

Erosion and sediment (E&S) control is an essential subcategory of the containment option and should be included and discussed as part of the RI/FS. The availability of the property limits to support conceptual E&S components such as sedimentation basins should be discussed.

Management of Hot Spots

Several evaluation criteria can be utilized to determine if hot spot management is applicable to the site conditions. The potential threats posed by the hot spot should be considered. Also, the accessibility of the hot spot for treatment should be considered, especially if the material is buried within the landfill. Finally, the volume of the hot spot material should be estimated.

The management of hot spots should be evaluated on a case-by-case basis because a wide range of remedial actions exist. Some general hot spot technologies that should be considered are summarized below.

- Excavation and on-site consolidation (within the existing landfill)
- On-site solidification/stabilization
- Off-site treatment and/or disposal
- Containment

If the hot spot is located within the landfill boundary, the presumptive remedy (i.e., capping) may address the situation. Hot spots located outside of the landfill boundary may be managed by consolidating the material within the landfill prior to final capping. If a preliminary decision has been made to address a hot spot by other means, such as solidification/stabilization, treatability testing may be required to facilitate remedial technology selection. Off-site remedies may be applicable to relatively small volumes of isolated waste materials.

Containment of hot spot areas outside of the landfill should also be considered as described in the previous section.

The excavation and consolidation technology consists of identifying, excavating, hauling and placing the hot spot material at a pre-determined location within the existing landfill limits. The materials may be dewatered (if needed) prior to excavation and consolidation. The hot spot area could be excavated via backhoe, hydraulic excavator, dredging via dragline, or pumping depending on the moisture content and condition of the material. Solidification/stabilization prior to excavation may be required to facilitate materials handling.

Potential solidification/stabilization technologies include:

• In-situ solidification/stabilization
• Ex-situ solidification/stabilization
• soil flushing

In-situ technologies involve the application of the technology to soils in-place. The application of any in-situ technology requires an understanding of the horizontal and vertical delineation of the treatment area and verification sampling. In-situ technologies may also require treatability or pilot study programs to evaluate the effectiveness and implementability of the technology. Ex-situ technologies require the delineation of the area, removal via excavation of the impacted material, treatment, verification of treatment, and the placement of treated materials back to the excavation area, or to other areas.

Soil flushing utilizes the injection of groundwater with the addition of surfactants to increase the mobility of the constituent of concern. The contaminants are then recovered in the groundwater by extraction wells or intercepter trenches for subsequent treatment.

Off-site treatment and/or disposal consists of the excavation and transportation of hot spot material to a permitted treatment, storage and disposal (TSD) facility. The TSD facility (depending on the nature of the impacted material) could consist of a municipal waste landfill, a chemical waste landfill, incinerator, or other facility. The excavated area would then be backfilled with general earthfill and revegetated.

Groundwater/Leachate Management

The feasibility evaluation of the groundwater/leachate management options is performed based upon a general understanding of the site specific geology and hydrogeology along with the depth of impacted groundwater. If the landfill contains a leachate collection system, the adequacy of that system should be evaluated initially. The major technologies that may be considered (at a minimum) for groundwater/leachate management (if necessary) include:

Groundwater Diversion/Isolation
- Slurry walls
- Sheet pilings
- Geosynthetic panels
- Grouting

Groundwater/Leachate Collection
- Interceptor trenches
- Well point network

Groundwater/Leachate Treatment
- On-site treatment followed by surface water discharge
- On-site pre-treatment followed by off-site treatment
- Off-site treatment
- In-situ remediation

Groundwater diversion/isolation technologies (vertical barriers) are utilized to contain, capture, or redirect groundwater flow. These systems are designed to prevent the migration of groundwater into the landfill mass and to prevent the migration of impacted groundwater (leachate) beyond the boundary of compliance. The barriers can be installed downgradient or upgradient of the landfill, or around the perimeter of the site. These vertical barriers would be "keyed into" the bedrock or other natural low permeability barrier, such as a clay layer, to contain the groundwater effectively. Vertical barriers are typically most cost effective and implementable at depths of less than 30 feet. Grouting may be utilized to seal voids in fractured bedrock. Grouting may be applied in conjunction with other vertical barrier technologies to obtain a competent barrier when fractured bedrock is the confining layer.

Groundwater/Leachate collection techniques involve manipulation of the groundwater by removing a plume, removing leachate from the source and/or adjusting (lowering) groundwater levels. This technology includes interceptor trenches and/or groundwater collection wells.

Interceptor trenches act as buried conduits to convey and collect groundwater/leachate as it flows into the trench. Trenches function as an infinite line of extraction wells and therefore may be utilized to collect impacted water or lower the groundwater table, in lieu of wells. Trenches are more effective than pumping wells in strata with low or variable hydraulic conductivity. Trenches are implementable at relatively shallow depths and their applicability may be limited by the cohesiveness of the surrounding soils.

Well point systems are suitable for aquifers where extraction is needed and the hydraulic properties allow an adequate barrier or collection scenario using wells. Typically, a series of well points are hydraulically connected in series to extract a cumulative volume of groundwater/leachate. Pumping wells can be effective where an aquifer is known to have a high hydraulic conductivity and the constituents of concern are present at depths greater than 30 feet.

Treatment technologies are designed to treat collected groundwater/leachate to meet discharge requirements. Discharge requirements are determined by the off-site treatment facility or surface water discharge requirements. Treatment options consist of on-site or off-site technologies. On-site pre-treatment/treatment technologies include biological, chemical and physical treatment processes. Off-site technologies consist of the use of the local Publicly Owned Treatment Works (POTW) or an industrial treatment facility. In-situ treatment technologies such as subsurface bioremediation may also be considered.

Landfill Gas Management

The applicable gas management system technologies for landfill gas consist of the following:

- Passive gas venting
- Active gas collection
- Landfill gas treatment

Passive systems are functional due to the natural pressure gradient (i.e. internal landfill pressure created due to landfill gas generation) or concentration gradient to convey the landfill gas to the atmosphere or a control system. A passive gas venting system consists of installing a series of vertically or horizontally oriented perforated collection pipes surrounded by granular material directly into the landfill contents to affect collection of gas. The collected gas is typically vented directly into the atmosphere. Passive gas venting systems can be designed such that they can be converted into an active collection system.

Active gas collection systems consist of vertically oriented venting systems and/or horizontal trench systems. The collection piping employs mechanical blowers or compressors to provide a pressure gradient to extract the landfill gas via a pipe header system. This network of extraction wells, trenches and pipes is designed to provide the capability of inducing negative pressure within the landfill. When the water saturated gas is extracted through the landfill, the decrease in pressure and temperature will result in the generation of condensate. The condensate is pumped through a force main to a collection vessel for subsequent treatment.

For passive systems, activated carbon canisters can be installed onto the vent for treatment of the gas. The most common technology used at landfill sites for the treatment of landfill gases collected via active systems is thermal treatment using ground flares. Enclosed ground flare systems consist of a refractory-lined flame enclosure or stack with a burner assembly at the base.

Several factors may be considered when determining whether to select an active or a passive gas management system including: (1) State or Federal requirements, (2) existing or potential off-site gas migration, (3) the gas generating potential of the landfill (including waste volume, waste age, and type of waste), (4) the existing or expected contaminants and/or odors of the gas, (5) the location of existing or planned structures and the potential threat of either an explosion or inhalation hazard, and (6) the final proposed usage of the Site. Interim regulations have been proposed by the EPA related to control of air emissions from landfills (Federal Register, AD-FRL-3780-9, May 30, 1991). Reportedly, when these rules become promulgated, they will apply to CERCLA-listed landfills.

Wetlands, Surface Water and Floodplain Management

Remediation of impacted wetlands and surface water may be accomplished by dredging the impacted sediments. However, these activities may result in wetland loss. The information contained in the risk assessment can be used to determine if the risk is significant enough to justify such activities. When natural wetlands are disturbed due to remediation and/or to perform construction activities (e.g. capping system installation), the wetlands may be restored or wetlands mitigation may be required. Surface water located directly adjacent to a landfill may also be relocated to reduce potential environmental impact and to facilitate remedial construction. These CERCLA activities do not require a United States Corps of Engineers (USCOE) permit (USCOE Regulatory Guidance Letter 94-02, August 17, 1994). However, compliance with the substantive permit requirements must be obtained. The potential for flooding to impact the final remedy should also be evaluated.

Institutional Controls and Site Management Procedures

The institutional controls and site management procedures that may be evaluated include:

- Access restrictions
- Local zoning ordinances and deed restrictions
- Site Monitoring following site closure
- Operation and maintenance (O&M)

Institutional controls and site management procedures consists of site-specific activities which may include maintaining access restrictions; securing deed restrictions; securing land-use restrictions or easements; and, performing monitoring activities necessary to verify the performance of the remedial action. Institutional controls are used to restrict site access or to ensure future site accesses are conducted in an approved manner. Institutional controls typically include future development restrictions. Institutional controls also include construction practice restrictions to ensure that post-remedial action conditions are not affected. Access restrictions would typically include installation of continuous chain-link fence around the site perimeter.

Monitoring of the implemented remedial action provides information needed to determine whether the remedial action objectives continue to be satisfied and to determine if additional or reduced action is necessary. Media-specific monitoring activities may include periodic sampling of air, surface water, soil, sediment, and groundwater. Technology specific monitoring requirements for the selected remedial action should be consistent with the overall monitoring requirements for remediation of the site. Technology specific monitoring may include monitoring of landfill gases, leachate monitoring, and periodic inspection of the containment system and related appurtenances as required.

O&M would include the activities required to sustain the remedial action during the post-remedial action period. Operational procedures would include the functional tasks required for the performance of any systems that are in-place. Maintenance procedures would include the routine inspections and follow-up actions necessary to maintain the remedial technologies.

REMEDIAL ALTERNATIVES

The remedial technologies that were ultimately selected in the FS are grouped into comprehensive remedial alternatives. These alternatives are then analyzed based upon effectiveness, implementability, and cost criteria (OERR Publication 9360.0-32, August, 1993). Following this analysis, a comparative analysis of the alternatives is performed in order to identify a recommended final alternative. This remedial alternative selection process is summarized below.

Analysis of Alternatives

The initial analysis involves evaluating the effectiveness and implementability of the alternatives. Effectiveness addresses the degree to which the alternative: meets the objective within the scope of the remedial action; complies with ARARs and other criteria, advisories and guidance; and, long term effectiveness and permanence. Implementability evaluation of alternatives includes an assessment of the technical feasibility, the administrative feasibility and, the availability of various services and materials.

Additionally, an economic evaluation is performed. Cost evaluations include an estimation of the capital cost (direct and indirect) and the annual Post-Removal Site Control (PRSC) cost. Specifically, the cost analysis consists of estimating the capital and PRSC costs for the technologies of each alternative, calculating the present worth for each alternative, performing a sensitivity analysis for changes in key parameters, and using the results in the comparative analysis. The cost analysis should be performed in accordance with the guidance contained in the documents titled *Remedial Action Costing Procedures Manual* (OSWER Publication EPA/600/8-87/049, October, 1987). Capital costs (direct and indirect) and PRSC costs are estimated using cost estimating information such as vendor information, Means Construction Cost Data (R.S. Means Company, 1995) and past experience at similar sites. The level of accuracy for this cost estimate is expected to range from +50 percent to -30 percent of the actual cost (OSWER Publication EPA/600/8-87/049, October, 1987).

Comparative Analysis of Alternatives

Following the individual analysis of alternatives, a comparative analysis is conducted to evaluate the relative performance of each alternative in relation to each analysis criteria (effectiveness, implementability and cost). The purpose of this analysis is to identify the advantages and disadvantages of each alternative relative to one another so that important tradeoffs can be identified.

Recommendation of Alternative

The results of the comparative analysis are used to identify the recommended alternative. The EPA is required to prepare a formal written response to the RI/FS. Ultimately, the EPA will determine the preferred action. This determination along with the RI/FS is placed in the Administrative Record. An Action Memorandum is prepared by the EPA as part of the EE/CA process. A Record of Decision (ROD) is prepared by the EPA in conjunction with the standard RI/FS process.

CONCLUSION

Based on the similar characteristics of landfill sites, streamlining the RI/FS is viable with respect to the characterization of site conditions and the subsequent development of remedial alternatives. Utilizing the presumptive remedy approach for landfill sites will promote focused data collection. Ultimately, the remedial design (RD) can be streamlined because technology-specific information can be collected as part of the RI.

REFERENCES

Dunn, I.S.; Anderson, L.R.; and F.W. Kiefer, (1980), *Fundamentals of Geotechnical Analysis*, John Wiley and Sons, New York.

Federal Register, AD-FRL-3780-9, RIN 2060-AC42, 40 CFR Parts 51, 52 and 60, *(Thursday, May 30, 1991), Standards of Performance for New Stationary Sources and Guidelines for Control of Existing Sources: Municipal Solid Waste Landfills; Proposed Rule, Guideline and Notice of Public Hearing.*

McDonald, M.G. and A.W. Harbaugh, (1988), *A modular three-dimensional finite-difference ground-water flow model. Techniques of Water-Resources Investigations of the U.S. Geological Survey, Book 6, Modeling Techniques,* Chap. A1.

OERR Publication 9360.0-32, EPA/540/R-93/057, (August 1993), *Guidance on Conducting Non-Time Critical Removal Actions Under CERCLA.*

ORD Publication EPA/625/4-91/025, (May 1991), *Design and Construction of RCRA/CERCLA Final Covers.*

OSWER Publication 9203.1-051, Volume 1, Numbers 1-5
- Volume 1, Number 2, (December 1992), *Early Action and Long-Term Action Under SACM-Interim Guidance.*

OSWER Publication 9355.0-47 FS, EPA/540/F-93/047, (September 1993), *Presumptive Remedies, Policies and Procedures.*

OSWER Publication 9355.0-49 FS, EPA/ 540/F-93/035, (September 1993), *Presumptive Remedy for CERCLA Municipal Landfill Sites.*

OSWER Publication 9355.3-01, EPA/540/G-89/004, (October 1988), *Guidance for Conducting Remedial Investigations and Feasibility Studies (RI/FS) Under CERCLA.*

OSWER Publication 9355.3-11, EPA/540/P-91/001, (February 1991), *Conducting Remedial Investigations/Feasibility Studies for CERCLA Municipal Landfill Sites.*

OSWER Publication 9355.3-11 FS, (September 1990), *Streamlining the RI/FS for CERCLA Municipal Landfill Sites,* Quick Reference Fact Sheet.

OSWER Publication EPA/600/8-87/049, (October 1987), *Remedial Action Costing Procedures Manual.*

USCOE Regulatory Guidance Letter 94-02, (August 17, 1994), by: John P. Elmore, P.E., *SUBJECT: Superfund Projects*.

R.S. Means Company, Inc., Kingston, MA (1995), *Means Heavy Construction Cost Data and Means Site Work Data*.

Shroeder, et. al., EPA/600/R-94/168a, (September 1994), *The Hydrologic Evaluation of Landfill Performance (HELP) Model, User's Guide for Volume 3*.

EARTHEN COVERS FOR SEMI-ARID AND ARID CLIMATES

by Craig H. Benson[1] and Milind V. Khire[2]

Abstract: This paper describes the use of earthen covers (capillary barriers or monolayer barriers) as alternatives to prescriptive covers designed as resistive barriers in semi-arid and arid environments. The principles behind capillary and monolayer barriers are described and the results of field studies are reviewed. Even though some of the field studies did not perform as well as intended, the data show that earthen covers can be effective barriers against percolation. Lessons learned from the field studies can also be used for guidance in future designs.

INTRODUCTION

In many regions of the United States, a prescriptive cover consisting of a barrier layer and a vegetated surface layer (Fig. 1) is employed or required by the governing regulatory agency when closing a landfill or capping an uncontrolled waste site. The barrier layer may be a layer of compacted clay, a geosynthetic clay liner, a geomembrane, or a combination of these layers. In some cases, an interlayer drain may be included to provide lateral drainage above the barrier layer. Regardless of the actual layering used, the mechanism primarily responsible for minimizing percolation is the hydraulic resistance afforded by the earthen or geosynthetic barrier layers (e.g., for clay, low saturated hydraulic conductivity, K_s). Thus, the prescriptive cover can be referred to as a "resistive barrier" (Benson et al. 1994, Schulz et al. 1989).

The prescriptive cover is generally effective in minimizing percolation and enhancing the efficiency of gas collection systems. Unfortunately, the prescriptive cover can be costly and thus can be impractical for rural landfills or remediation of uncontrolled facilities for which viable responsible parties do not exist. However, in semi-arid and arid regions, economical alternative covers constructed solely from earthen materials can prove to be as effective in limiting percolation as the prescriptive cover. Although these covers may not be as effective in limiting gas migration, the volume of gas generated at these sites is sufficiently low such that management of landfill gas is not problematic.

This paper describes the principles behind two types of alternative earthen covers, capillary barriers and monolayer soil barriers, and contains a review of field

---

[1] Assoc. Prof., Dept. of Civil and Environ. Eng., Univ. of Wisconsin, Madison, WI 53706
[2] Grad. Res. Asst., Dept. of Civil and Environ. Eng., Univ. of Wisconsin, Madison, WI 53706

studies that have been conducted to evaluate the performance of alternative covers. Some, but not all of the studies have been conducted in semi-arid or arid climates. However, from each study lessons can be learned that can be applied when considering, designing, or evaluating an alternative cover for a semi-arid or arid environment.

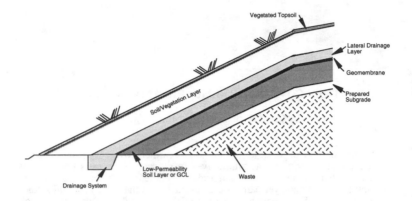

Fig. 1 Prescriptive Final Cover.

WATER BALANCE OF EARTHEN COVERS

Alternative earthen covers generally exploit the unique characteristics of unsaturated flow, the storage capacity of fine-grained soils, and the natural capacity of plants to remove water entering the cover during wet periods. These factors are linked by the water balance, which accounts for movement of water into, within, and out of a final cover. The water balance consists of precipitation (P) in the form of snow, rain, or ice; overland flow (O); soil water storage (S); evaporation from the surface (E); transpiration by vegetation (T); lateral drainage (L); and percolation from the base (P_r) (Fig. 2). In some cases, evaporation and transpiration are combined as evapotranspiration (E_t).

In algebraic form, the water balance can be described by the following equation:

$$P_r = P - O - S - E - T - L \qquad (1)$$

The form of Eq. 1 indicates that percolation can be minimized by enhancing overland flow, soil water storage, evaporation, transpiration, or lateral drainage. Of these factors, soil water storage and lateral drainage are easier to optimize during design. For example, soil water storage can be increased by selecting surficial soils containing a greater percentage of fines or by increasing the thickness of the cover, whereas lateral drainage can be increased by adding a wicking layer (Yeh et al. 1994) or employing a layer with anisotropic hydraulic conductivity (Stormont 1995). Evaporation can be enhanced by selecting soils having unsaturated hydraulic conductivity (K_ψ) that changes gradually with matric suction (ψ) (Khire et al. 1995). Fine-grained soils typically have this type of unsaturated hydraulic conductivity function (Hillel 1980, Meerdink 1994). Transpiration can be enhanced by careful selection of vegetation and manipulating the extent and density of the plant canopy (Rockhold et al. 1995)

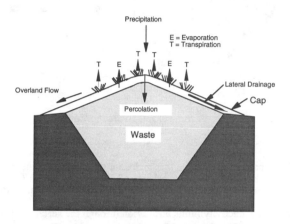

Fig. 2. Water Balance of Landfill Final Cover.

Design of an alternative earthen cover for a semi-arid or arid environment entails manipulating soil water storage capacity, lateral drainage, transpiration, and evaporation such that an acceptable value for percolation occurs in a worst case design year (e.g., year where rainfall or snowfall is abnormally high and temperature and solar radiation are abnormally low). In particular, the cover is designed such that it has adequate capacity to store or divert water that infiltrates during late fall and winter and sufficient vegetation and evaporative potential to remove the stored water during spring, summer, and early fall. Two types of earthen covers that are designed on this principle are capillary barriers and monolayer soil barriers. In some cases, these barriers are used in conjunction with a resistive barrier.

Capillary Barriers

In their simplest form, capillary barriers employ a fine-grained layer over a coarse-grained layer (Fig. 3a). Flow across the interface of these layers is restricted under unsaturated conditions because the unsaturated hydraulic conductivity of the coarse-grained layer is much lower than the unsaturated hydraulic conductivity of the fine-grained layer (Fig. 3b) (Hillel 1980). Thus, the fine-grained soil can store or divert water that infiltrates into the cover, and yet flow into the coarse-grained layer is restricted. More elaborate designs employing multiple layers having contrasting grain size are also possible. These covers employ the capillary barrier principle to divert infiltrating water via lateral flow while ensuring that deep percolation does not occur (Nyhan et al. 1993, Yeh et al. 1994).

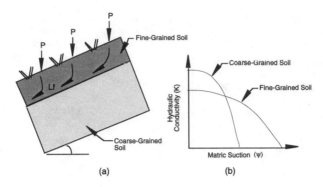

Fig. 3. Schematic of Capillary Barrier (a) and Hydraulic Conductivity Functions (b).

Monolayer Barriers

Monolayer barriers are covers that include a thick layer of fine-grained soil. Monolayer barriers exploit two characteristics of fine-grained soils: (1) their large soil water storage capacity when unsaturated and (2) their low saturated hydraulic conductivity relative to coarse-grained soils (Morrison-Knudsen 1993). Their low saturated hydraulic conductivity limits infiltration through the surface during rainfall or snow melt. Their high water storage capacity provides the capability to store water that does infiltrate until it can later be removed by evapotranspiration. The barrier must be sufficiently thick, however, such that changes in water content (θ) do not occur near its base; i.e., all changes in soil water storage occur in the upper portion of the barrier (Fig. 4). Otherwise, percolation will occur. The necessary thickness is a function of the type of precipitation received, the unsaturated hydraulic properties of the soil, and the rate at which water can be removed by evapotranspiration. Monolayer barriers are constructed from silty sands, silts, and clayey silts, and are cost-effective when large quantities of fine-grained soil requiring little processing is available on site.

FIELD STUDIES

Sophisticated water balance computer models (e.g., UNSAT-H described by Fayer and Jones 1990) can be effective tools for designing earthen coves (Fayer et al. 1992, Khire et al. 1994b). However, verification of the existing models has been limited and sometimes is conflicting (Khire et al. 1994b, Rockhold et al 1995). In addition, predictions made with these models are very sensitive to the unsaturated hydraulic properties used as input (Khire et al. 1995), which are often not known a priori. Consequently, acceptance of an earthen alternative final cover commonly includes a large-scale field test or extensive monitoring.

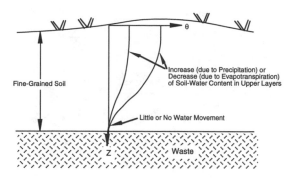

Fig. 4. Monolayer Final Cover.

Capillary Barriers

Melchoir et al. (1994)

Melchoir et al. (1994) constructed five large-scale test sections (Fig. 5) at a landfill near Hamburg, Germany (average annual precipitation is 800 mm). Each test section was 10 m x 50 m in areal extent, and the slope varied from 4% to 20%. All test sections were designed as multi-layered systems with a combination of surficial topsoil, drainage layers, capillary barriers, and resistive barriers. Overland flow, lateral flow, and percolation were measured. Neutron probes and tensiometers were installed to measure soil water content and matric suction. A weather station recorded on-site weather data.

The data indicate that capillary barriers, when used in conjunction with clay resistive barriers, can reduce percolation to virtually zero (Fig. 5), even in relatively wet climates such as Hamburg. Overland flow accounted for only 2% of precipitation, even on steep slopes; the remaining water evapotranspired (64% of precipitation) or infiltrated through the 750 mm surface layer. The lateral drainage layer proved very effective in removing a high percentage of infiltrated water (30% to 40% of precipitation).

Montgomery and Parsons (1990)

Montgomery and Parsons (1990) measured percolation and overland flow for three 12.2 m x 6.1 m test sections constructed at Omega Hills Landfill near Milwaukee, Wisconsin (average annual precipitation = 800 mm). Two test sections were designed as prescriptive covers containing resistive barriers (compacted clay layers 1200 mm thick). One of the test sections (RB1) had a 150-mm-thick top soil layer above the clay layer, whereas the other top soil layer was 450 mm thick (RB2) (Fig. 6). The third test section (CB) was designed as a capillary barrier underlain by a resistive barrier. The test section consisted of a compacted clay layer 610 mm thick, overlain by 300 mm of sand, overlain by another 610 mm of compacted clay. A layer of topsoil 150 mm thick was also placed on the surface. All of the test sections had a slope of 33%.

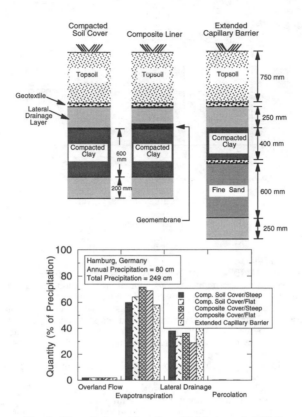

Fig. 5. Water Balance Data from Melchoir et al. (1994)

Data were collected for 4 years. Lysimeters were used for measuring percolation, diversion berms were used for collecting overland flow, and water contents were measured with neutron probes. Meteorologic data consisted of rain, air temperature, and relative humidity. A summary of the data is contained in Fig. 6.

Percolation observed from the two resistive barriers was similar, although slightly more percolation emanated from the resistive barrier with a thicker topsoil layer (4.8% of precipitation, compared to 3.6%). Apparently, the additional evapotranspiration expected from the thicker topsoil layer was not realized. The capillary barrier also did not perform as designed. The upper clay layer became extensively cracked and allowed rapid infiltration of water into the sand layer; however, the sand layer removed a large fraction of the infiltrating water via lateral flow. The lower clay layer remained intact and percolation through it continued at a relatively steady rate (4.6% of precipitation).

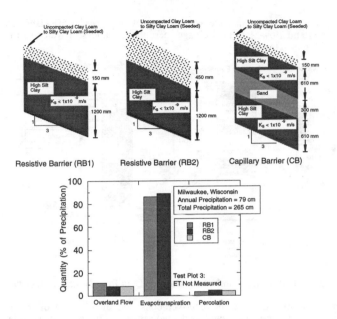

Fig. 6. Test Sections and Water Balance Data from Montgomery and Parsons (1990)

Nyhan et al. (1990)

Nyhan et al. (1990) compared the hydrology of a prescriptive cover to that of an alternative capillary barrier at a site in Los Alamos, New Mexico. Two test sections 3 m x 10.7 m in plan view were constructed. The test section simulating a prescriptive cover consisted of a 200-mm-thick surface layer comprised of sandy loam underlain by 1080 mm of crushed tuff backfill. The capillary barrier consisted of 710-mm-thick surface layer of sandy loam underlain by 460 mm of gravel (5-10 mm diameter). The gravel was underlain by a 910-mm-thick biota barrier consisting of cobbles. Crushed tuff backfill was placed below the biota barrier.

The data indicate that the alternative capillary barrier was more effective in limiting percolation than the prescriptive cover (Fig. 7); percolation produced by the prescriptive cover was four times higher than percolation from the capillary barrier. Furthermore, evapotranspiration from the capillary barrier (96% of precipitation) was larger than evapotranspiration from the prescriptive cover (88% of precipitation). It was also observed that plant species different from those originally seeded eventually covered the test sections.

Nyhan et al. (1993)

Nyhan et al. (1993) studied the water balance of four landfill cover test sections constructed at Los Alamos National Laboratory. The test sections consisted of the following cover designs: conventional prescriptive cover, EPA-recommended prescriptive cover, loam capillary barrier, and a clay-loam capillary barrier (Fig. 8).

The test sections were 1 m wide x 10 m long, with slopes of 5%, 10%, 15%, and 25%. Overland flow, lateral drainage, percolation, and soil water storage were measured for each test section.

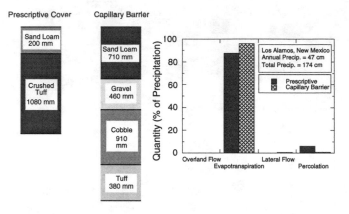

Fig. 7. Test Sections and Water Balance Data from Nyhan et al. (1990)

The conventional prescriptive cover consisted of 150 mm of loam topsoil comprised of organic matter, sand, and aged sawdust. The topsoil was underlain by 760 mm of crushed tuff. The EPA-recommended design contained 610 mm of loam topsoil placed on top of a layer of medium sand 300 mm thick. The bottom layer consisted of a 600-mm-thick layer of clay tuff having saturated hydraulic conductivity less than 1×10^{-9} m/s. Both capillary barrier designs employed 760 mm of fine sand as a wicking layer for lateral drainage. The loam capillary barrier had 610 mm of loam topsoil overlying the fine sand, whereas the clay-loam capillary barrier consisted of 610 mm of clay loam overlying the fine sand. Both test sections had 300 mm of gravel and 200 mm of coarse sand underlying the fine sand (Fig. 8).

The test sections were monitored for 15 months. A summary of the data is shown in Fig. 8. Analysis of the data showed that all of the test sections produced percolation, and that the EPA-recommended resistive barrier and the loam capillary barrier produced the largest quantity of percolation (8.5% and 7.4% of precipitation, respectively). The clay-loam capillary barrier produced the least amount of lateral flow (0.7% of precipitation), the greatest amount of overland flow (6.2% of precipitation), and the least amount of percolation (0.7%). The superior performance of the clay-loam capillary barrier is attributed to the lower saturated hydraulic conductivity afforded by the clay-loam surface layer.

Wing and Gee (1994)

Wing and Gee (1994) constructed test lysimeters simulating capillary barriers at the U.S. Dept. of Energy's Hanford site in Richland, Washington. Rockhold et al. (1995) provide a detailed description of the lysimeters and the data that were collected. Wing and Gee (1994) report that the capillary barriers were effective in controlling downward movement of water. They also discovered that capillary barriers can be ineffective in regions with significant snow cover. Water from snow melt overwhelms the storage capacity of the fine-grained surface layer, resulting in significant infiltration into the coarse-grained layer and percolation into underlying waste. Wing and Gee (1994) also report vegetation enhances the performance of capillary barriers because of the additional water removed by transpiration.

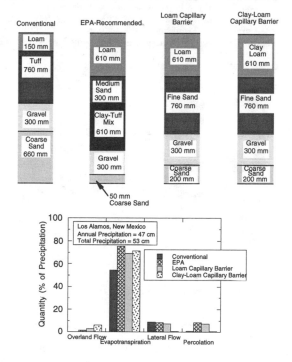

Fig. 8. Test Sections and Water Balance Data from Nyhan et al. (1993)

Hakonson et al. (1994)

Four landfill cover designs (conventional prescriptive cover, "modified-RCRA" prescriptive cover, and two capillary barriers) constructed at Hill Air Force Base, Utah were evaluated by Hakonson et al. (1994). The test sections were constructed in modular swimming pools having dimensions (plan view) of 5 m x 10 m. All sections had a 4% slope, and were instrumented to monitor precipitation, air temperature, overland flow, soil water storage, soil temperature, percolation, lateral drainage, and erosion. Soil water content was measured at various depths using a neutron moisture gauge. Overland flow and sediments were collected in a large tank. Percolation was measured by directing it into a large underground caisson, where a tipping bucket was used to measure the flow rate.

The conventional prescriptive cover consisted of a 900-mm-thick sandy loam (K_s = 2.8 x 10^{-6} m/s) overlying a 300-mm-thick gravel drainage layer (Fig. 9), which in essence is a simple, two-layer capillary barrier. The "modified-RCRA" prescriptive cover consisted of 1200 mm of sandy loam topsoil overlying a sand drainage layer

(300 mm thick), and a layer of clay loam amended with bentonite (600 mm thick). The saturated hydraulic conductivity of the loam-bentonite layer was 3.4×10^{-8} m/s. The test sections designed as capillary barriers had the same layering, except one was seeded with grass, whereas the other was seeded with grass and shrubs to enhance evapotranspiration. Both capillary barriers had 1500 mm of sandy loam topsoil over 300 mm of washed gravel (~ 10 mm diameter). A non-woven geotextile was placed between the layers to limit migration of sand and fines into the gravel so that a sharp capillary interface could be preserved. A thin gravel cover was placed on each capillary barrier test section to reduce erosion.

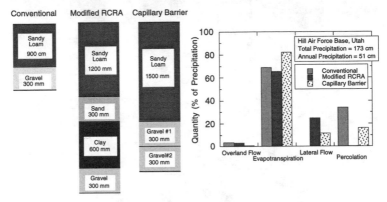

Fig. 9 Test Sections and Water Balance Data from Hankonson et al. (1994)

The test sections were studied for 46 months, during which 1730 mm of precipitation (28% snow) was received. The long-term average precipitation is 510 mm/yr. A summary of the data is shown in Fig. 9. Percolation from the modified-RCRA prescriptive cover was the least (0.006% of precipitation), whereas percolation from the conventional prescriptive cover was greatest (24% of precipitation). Both capillary barriers produced percolation that was approximately 15% of the precipitation. Percolation from the capillary barriers was large because snow accumulated on the test sections, which overwhelmed their capacity for storage and diversion.

The greatest overland flow was obtained for the conventional cover (3.4% of precipitation), whereas the least was obtained for the capillary barrier with grass (0.8% of precipitation). Overland flow for the capillary barrier with grass and shrubs was 1.3% of precipitation, and for the RCRA design it was 5.5% of precipitation.

Evapotranspiration was largest for the capillary barrier with grass (84% of precipitation) and smallest for the RCRA design (65% of precipitation). Surprisingly, the shrubs added to the capillary barrier had no beneficial influence on evapotranspiration (83% of precipitation).

Lateral flow was greatest for the RCRA design (25% of precipitation). No lateral flow was recorded for the conventional prescriptive cover. Lateral flow also occurred in the capillary barriers, being 11% for the cover vegetated only with grass only and 7% for the cover vegetated with grass and shrubs. However, Hakonson et al. (1994) do not describe through which layers lateral flow occurred.

Khire et al. (1994b)

Khire et al. (1994b) describe the water balance of two final cover test sections (resistive and capillary barriers) constructed adjacent to each other in a semi-arid climate (East Wenatchee, Washington; annual precipitation = 230 mm). The two test sections (Fig. 10a) are 30 m x 30 m in areal extent and are located on a landfill side slope. The prescriptive resistive barrier is constructed with a compacted silty clay barrier 600 mm thick and a silty clay surface layer 150 mm thick. The capillary barrier has a sand layer 750 mm thick overlain by a surface layer of silt 150 mm thick. The capillary barrier was constructed without a geotextile separator between the fine- and coarse-grained layers. Examination of the interface between the layers has shown, however, that a distinct interface exists. Benson et al. (1994) provide a detailed description of how the test sections were constructed and instrumented.

Khire et al. (1994b) report that the capillary barrier has been more effective than the resistive barrier in restricting percolation (Fig. 10b). Percolation from the capillary barrier has been 0.6% of precipitation, whereas percolation from the resistive barrier has been 4.4% of precipitation. Overland flow and evapotranspiration from both test sections have been similar, being 15% and 65% of precipitation, respectively.

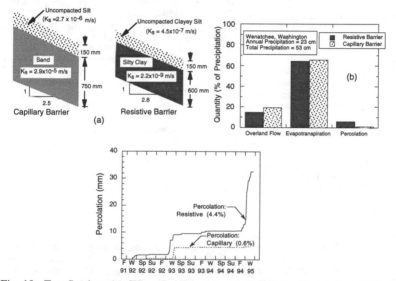

Fig. 10   Test Sections (a), Water Balance (b), and Cumulative Percolation (c) from Khire et al. (1994)

Khire et al. (1994b) report findings regarding snow accumulation similar to those by Wing and Gee (1994); that is, the capillary barrier can be ineffective when snow accumulates on the cover and the subsequent melt overwhelms the storage and diversion capacity of the fine-grained layer. For example, approximately the same amount of precipitation was recorded during the winters of 1992-93 and 1993-94. However, in 1992-93, 1.68m of snow accumulated on the test sections. When the snow melted in late February 1993, the fine-grained layer became saturated and rapid flow occurred through the coarse-grained layer. Subsequently, a large pulse of percolation occurred (Fig. 10c). In contrast, only 0.09 m of snow accumulated in 1994-95. When the snow melted in early March 1995, the fine-grained layer became nearly saturated, but the quantity of infiltration did not overwhelm the storage capacity of the fine-grained layer. Consequently, percolation from the cover was nearly imperceptible (Fig. 10c).

The key difference between these two winters is that in 1992-93, most of the winter precipitation was stored as a thick snow canopy and then applied to the test section as water in a short period when it melted. As a result, a large fraction of the precipitation infiltrated, and because the test section was covered with snow, very little water evaporated or transpired. However, in 1994-95, little precipitation was stored on the surface as snow. Instead, water was applied to test section as light rains, or snows that quickly melted. Thus, most of the water applied during 1994-95 evaporated, was diverted as lateral flow, or transpired between precipitation events.

Thus, when designing the fine-grained layer for a capillary barrier in a region where snow accumulates, it is essential to account for accumulation of snow and how water in the snow is applied to the surface layer. If most of the winter precipitation accumulates as snow, and the snow melts in a short period during a spring melt, then the fine-grained layer must be designed such that it has sufficient capacity to store a large fraction of the total precipitation received during winter.

Problems with desiccation cracking and biota intrusion in the prescriptive resistive barrier have also been reported by Khire et al. (1994b). Large vertical desiccation cracks have been observed when excavating test pits, and animal burrows are evident on the surface. In contrast, the capillary barrier is devoid of such features. Apparently, the cracks and biota intrusions are pathways for preferential flow. During intense storms in the summer months, pulses of percolation have been measured whereas monitoring has shown no change in water content except at shallow depths. No percolation has emanated from the capillary barrier during these same storms.

Khire et al. (1994a,b) also report that a better stand of vegetation generally exists on the resistive barrier, which typically has higher water content than the capillary barrier. Also, the predominant vegetation on both test sections consists of a mixture of native species rather than the species that were seeded. This suggests that it may be difficult to implement a cover seeded with vegetation selected to maximize evapotranspiration unless the species is native to the area.

Monolayer Barriers

Morrison-Knudson (1993, 1994)
Morrison-Knudson (1993, 1994) describe a monitoring program used to assess the performance of a monolayer barrier proposed as a final cover over contaminated soil at Rocky Mountain Arsenal in Commerce City, Colorado. The barrier, which is constructed over trenches containing hazardous constituents, consists of a vegetated layer of topsoil underlain by a compacted layer of fine-grained soil. The minimum combined thickness of the topsoil and fine-grained soil layers is 1.2 m (Fig. 11).

Six nests of water content sensors are being used to monitor water contents at various depths. The locations were selected to monitor water contents where different slopes and thicknesses of the cover exist. Monitoring began in May 1992 and continued through April 1994. Analysis of the data has shown that most of the changes in water content occur in the upper 0.6 m of the cover. However, gradual changes in water content have been observed in the deepest probes (1.2 m).

Unfortunately, no lysimeters were installed beneath the cover. Thus, no assessment can be made regarding percolation from the base and the true performance of the cover. Also, no test pits have been excavated to determine if desiccation cracks have formed in the cover.

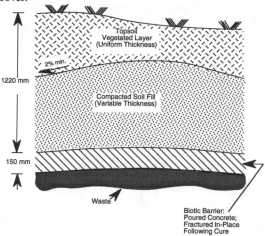

Fig. 11. Monolayer Barrier Used at Rocky Mountain Arsenal.

GeoLogic Associates (1993)

GeoLogic Associates (1993) describe a monitoring program conducted to assess the performance of a thick soil barrier used as a final cover for a landfill in Southern California. The barrier is 2 m thick and is constructed from a clayey silt. The barrier is instrumented with a network of probes to monitor water content and meteorologic conditions.

To date, water movement has been limited to the upper 0.6 m of soil. No changes in water content have been observed for the deeper probes (deepest = 2 m). The data indicate that the water content of the upper soil layers increases rapidly after rainfall and then gradually decreases as water is removed by evapotranspiration.

LESSONS LEARNED FROM FIELD STUDIES

Earthen Covers Can Be Effective

Field studies have shown that capillary barriers in their simplest form (two layer) are effective final covers in semi-arid and arid regions. In addition, more complex designs having more than two layers have been found to be effective in humid regions. Nyhan et al. (1990) and Khire et al. (1994a,b) have shown that capillary barriers can be more effective than prescriptive final covers constructed from earthen materials. For example, the prescriptive resistive barrier described by Khire et al. (1994b) has generated six times as much percolation than the adjacent capillary barrier. Lane et al. (1992) also report that the capillary barrier described by

Khire et al. (1994b) was easier to construct and less costly than the prescriptive resistive barrier.

Sufficient data regarding the performance of monolayer barriers have not yet been gathered. Nevertheless, the study by Geologic Associates (1993) suggests that monolayer barriers may be effective in some regions. Field tests including large-scale measurements of percolation are needed before definitive conclusions regarding monolayer barriers can be drawn.

Capillary Barriers: Hydraulic Conductivity of the Surface Layer

Capillary barriers perform better if the saturated hydraulic conductivity of the surface layer is lower. Nyhan et al. (1993) report that percolation from a capillary barrier constructed with a clay-loam surface layer had 11 times less percolation than an identical capillary barrier constructed with a loam surface layer. The capillary barrier constructed with a clay-loam surface layer also shed more water as overland flow.

Capillary Barriers: Storage Capacity of the Surface Layer

The surface layer of a capillary barrier must be designed to have adequate storage capacity such that significant flow into the coarse-grained layer does not occur. This is particularly important in regions where snow accumulates, because large pulses of percolation can occur if the surface layer has inadequate capacity to store water that infiltrates during snow melt (Wing and Gee 1994, Khire et al. 1994a,b). In regions where snow accumulates, the critical storage capacity should be defined using the storage capacity required in the spring when snow melt occurs. Thus, the design year for regions where snow accumulates should correspond to the year with maximum snow accumulation.

Desiccation Cracking and Biota Intrusion

The surface layer of a capillary barrier must not crack or the effectiveness of the barrier will be compromised. Thus, caution must be exercised when selecting soils for the surface layer. Suitable soils are likely to be clayey silts, silty sands, and some sandy clays. Clay-rich soils should be avoided if possible, because clays shrink and crack when dried (Montgomery and Parsons 1990). If clays are used, they should be placed at low water contents to minimize desiccation cracking (Kleppe and Olson 1984, Daniel and Wu 1993). Furthermore, if significant potential exists for the surface layer to shrink and crack on drying, then a three layer capillary barrier may prove to be a more effective alternative (Yeh et al. 1994).

Barriers constructed with thick layers of fine-grained soil must be protected from intrusion by biota. Burrows or tunnels created by animals or deep root holes can become preferential flow paths (Khire et al. 1994b). Thus, biota barriers should be included in any design where preferential flow through a fine-grained layer will compromise performance of the cover.

Vegetation

Field data show that test sections designed to yield greater evapotranspiration do not necessarily perform as intended (Montgomery and Parsons 1990, Hakonson et al. 1994). Furthermore, dense vegetation or species that are not native can be difficult to sustain on capillary barriers (Khire et al. 1994a,b). Thus, designers should be cautious regarding the benefits that can actually be accrued through enhanced vegetation.

SUMMARY

Field data from several sites have confirmed that earthen final covers designed as capillary barriers can be effective final covers in semi-arid and arid regions. In some cases, where the soils needed for construction are available on site, earthen covers of this type can be less costly than prescriptive covers designed as resistive barriers. Less data are available regarding monolayer barriers, but some of the data collected to date suggest that these covers may also be effective. More data must be collected, however, before a definitive conclusion can be drawn.

The field data also indicate that several factors have a strong influence on the performance of alternative earthen covers. For capillary barriers, these factors include saturated hydraulic conductivity and storage capacity of the fine-grained layer, lateral diversion capacity, and resistance to desiccation cracking. Desiccation cracking is also an issue with monolayer barriers, as are preferential pathways created by intrusion of biota. Finally, the field data also suggest that additional evapotranspiration expected due to enhanced vegetation may not be realized. Considering these factors during design will likely result in an earthen barrier that is more likely to perform as expected.

ACKNOWLEDGMENT

Financial support for a portion of the work described in this paper has been provided by the National Science Foundation and WMX, Inc. Support from NSF was provided through grant no. CMS-9157116. This paper has not been reviewed by NSF or WMX and no endorsement should be assumed.

REFERENCES

Benson, C., Bosscher, P., Lane, D., and Pliska, R. (1994), "Monitoring System for Hydrologic Evaluation of Landfill Final Covers," *Geotechnical Testing Journal*, ASTM, Vol. 17, No. 2, pp. 138-149.

Daniel, D. and Wu, Y. (1993), "Compacted Clay Liners and Covers for Arid Sites," *J. Geotech. Engrg.*, Vol. 119, No. 2., pp. 223-237.

Fayer, M. and T. Jones (1990), "Unsaturated Soil-Water and Heat Flow Model, Ver. 2.0," Pacific Northwest Laboratory, Richland, Washington.

Fayer, M., Rockhold, M., and Campbell, M. (1992), "Hydrologic Modeling of Protective Barriers: Comparison of Field Data and Simulation Results," *Soil Science Society of America Journal*, Vol. 56, pp. 690-700.

GeoLogic Associates (1993), "Evaluation of Unsaturated Fluid Flow, Coastal Sage Scrub Habitat Area, Coyote Canyon Final Cover System, Orange County,

California," report prepared by GeoLogic Associates for Orange County Integrated Waste Management Department.

Hakonson, T., Bostick, K., Trulillo, G., Manies, K., Warren, R., Lane, L., Kent, J., and Wilson, W. (1994), "Hydrologic Evaluation of Four Landfill Cover Designs at Hill Air Force Base, Utah," Dept. of Energy Mixed Waste Landfill Integrated Demonstration Sandia National Laboratory, LAUR-93-4469.

Hillel, D. (1980), *Fundamentals of Soil Physics*, Academic Press, Inc.

Khire, M., Meerdink, J., Benson, C., and Bosscher, P. (1995), "Unsaturated Hydraulic Conductivity and Water Balance Predictions for Earthen Landfill Final Covers," *Soil Suction in Geotechnical Engineering Practice,* Geotechnical Special Publication, ASCE, in press, to appear Sept. 1995.

Khire, M., Benson, C., Bosscher, P., and Pliska, R. (1994a), "Field-Scale Comparison of Capillary and Resistive Landfill Covers in an Arid Climate," *Fourteenth Annual Hydrology Days Conference*, Fort Collins, CO, pp. 195-209.

Khire, M., Benson, C., and Bosscher, P. (1994b), "Final Cover Hydrologic Evaluation-Phase III," Environmental Geotechnics Report 94-4, Dept. of Civil and Environmental Engineering, University of Wisconsin-Madison.

Kleppe, J. and Olson, R. (1984), "Desiccation Cracking of Soil Barriers," *Hydraulic Barriers in Soil and Rock*, ASTM STP 874, American Society for Testing and Materials, Philadelphia, pp. 263-275.

Lane, D., Benson, C., and P. Bosscher (1992), "Hydrologic Observations and Modeling Assessments of Landfill Covers-Final Report-Phase I," Environmental Geotechnics Report 92-10, Dept. of Civil and Environmental Engineering, University of Wisconsin-Madison.

Meerdink, J. (1994), "Unsaturated Hydraulic Conductivity of Barrier Soils Used for Final Covers," MS Thesis, Dept. of Civil and Environmental Engineering, University of Wisconsin-Madison.

Melchoir, S., Berger, K., Vielhaber, B., and Miehlich, G. (1994), "Multilayered Landfill Covers: Field Data on the Water Balance and Liner Performance," *Proceedings of the 33rd Hanford Symposium on Health and Environment*, Pasco, WA, Nov. 7-11, pp. 411-425.

Montgomery, R. and Parsons, L. (1990), "The Omega Hills Cover Test Plot Study: Fourth Year Data Summary," *Proceedings of the 22nd Mid-Atlantic Industrial Waste Conference*, Drexel University, July 24-27, 1990.

Morrison-Knudsen (1993), "White Paper, Implementation of Soil/Vegetative Covers for Final Remediation of the Rocky Mountain Arsenal," prepared by Morrison-Knudsen Corporation, Denver, CO, for Shell Oil Company, Dec. 1993.

Morrison-Knudsen (1994), "Annual Monitoring Report for Other Contamination Sources, Interim Response Action, Shell Section 36 Trenches Containment

System," prepared by Morrison-Knudsen Corporation, Denver, CO, for Shell Oil Company, Denver, CO, Aug. 1994

Nyhan, J., Langhorst, G., Martin, C., Martinez, J., and Schofield, T. (1993), " Hydrologic Studies of Multilayered Landfill Closure of Waste Landfills at Los Alamos, *Proceedings of 1993 DOE Environmental Remediation Conference "ER 93"*, Oct. 1993, Augusta, GA, U.S.A.

Nyhan, J., Hakonson, T., and Drennon, B. (1990),"A Water Balance Study of Two Landfill Cover Designs for Semiarid Regions," *Journal of Environmental Quality*, Vol. 19, pp. 281-288.

Rockhold, M, Fayer, M., Kincaid, C., and G. Gee (1995),"Estimation of Natural Ground Water Recharge for the Performance Assessment of a Low-Level Waste Disposal Facility at the Hanford Site," PNL-10508, Pacific Northwest Laboratory, Richland, Washington.

Schulz, R.K., Robert, R.W., and E. O'Donnell (1989), "Control of Water Infiltration Into Near Surface LLW Disposal Units, Annual Report," prepared for the U.S. Nuclear Regulatory Commission, NUREG/CR-4918, Vol. 3.

Stormont, J.C. (1995), "The Effect of Constant Anisotropy on Capillary Barrier Performance," *Water Resources Research*, Vol. 31, No. 3, pp. 783-786.

Yeh, T., Guzman, A., Srivastava, R., and Gagnard, P. (1994),"Numerical Simulation of the Wicking Effect in Liner Systems," *Ground Water*, 32(1), pp. 2-11.

Wing, R. and Gee, G. (1994), "Quest for the Perfect Cap," *Civil Engineering*, October 1994, pp. 38-41.

DESIGN OF MSW LANDFILL FINAL COVER SYSTEMS

Majdi A. Othman[1], Rudolph Bonaparte[1], Beth A. Gross[2],
and Gary R. Schmertmann[1], Members, ASCE

Abstract

This paper summarizes the current state of practice regarding the design of "conventional" final cover systems for municipal solid waste (MSW) landfills in the United States. The paper provides brief descriptions of design methods and practices which are commonly used by the general engineering community. Where applicable, the advantages and disadvantages of using more sophisticated methods are also discussed. The major design aspects considered relate to: (i) flow of water in and through the final cover system; (ii) impacts of waste settlement on the performance of final cover system components; (iii) static and dynamic cover system stability; and (iv) surface-water management.

Introduction

Overview: The purpose of this paper is to review methods and practices commonly used by the general engineering community in the design of final cover systems for MSW landfills. Final cover systems form one component of the integrated group of engineered systems used at landfills to achieve environmentally safe land disposal of MSW. Other components include liner systems, daily and intermediate cover systems, leachate collection and removal systems, gas collection and removal systems, and surface-water management systems. The general layout of these systems (excluding the intermediate cover system) at a landfill is shown in Figure 1.

---

[1]GeoSyntec Consultants, 1100 Lake Hearn Drive, NE, Atlanta, GA 30342.
[2]GeoSyntec Consultants, 1004 East 43rd Street, Austin, TX 78751.

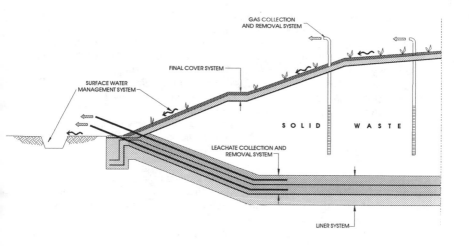

Figure 1.   General layout of engineered systems used for liquid and gas containment/collection at MSW landfills.

The principal functions of a landfill final cover system are:

- minimize water and air infiltration into the landfill;

- minimize gas migration out of the landfill;

- serve as a system for control of odors, disease vectors, and other nuisances; and

- serve as a component of the landfill surface-water management system.

The focus of this paper is on "conventional" final cover systems consisting of a series of soil and geosynthetic layers. Typically, these systems contain hydraulic barrier layers overlain by drainage and surface layers. Alternative final cover systems, including, for instance, monolithic covers, covers incorporating capillary barriers, and covers incorporating cobble surface layers, are applicable to limited specific design applications, particularly certain applications in arid environments. In these environments, it is difficult to maintain the required moisture content of soil barrier layers during construction, the soil barrier layers are susceptible to desiccation cracking, the surface vegetation is stressed by lack of available water, and the surface layer is susceptible to erosion. The authors are aware of field studies to evaluate the performance of alternative cover systems. Studies are currently underway at a number of locations, including sites near El Paso, Texas, Albuquerque, New Mexico, and Richland, Washington. Due to page-limit constraints, these alternative cover systems are not addressed in this paper.

Federal Regulations: The U.S. Environmental Protection Agency (USEPA) minimum requirements for final cover systems for MSW landfills are contained in 40 CFR §258.60(a). These regulations were promulgated under the authority of "Subtitle D" of the Resource Conservation and Recovery Act (RCRA) and subsequently clarified by USEPA in 57 FR 28626-28628 (USEPA, 1992). According to this clarification, the final cover system is required to have a hydraulic conductivity that is less than or equal to that of any underlying liner system or natural subsoils. The purpose of this requirement is to prevent what USEPA calls the "bathtub" effect wherein infiltration into the landfill exceeds exfiltration through the liner system, causing the accumulation of liquid in the facility. The hydraulic conductivity must also be no greater than 10^{-5} cm/s.

The USEPA (1992) clarification on the minimum requirements for MSW landfill final cover systems is illustrated in Figure 2 for: (i) unlined landfills, constructed prior to the effective date of Subtitle D regulations (Figure 2(a)); (ii) landfills with a compacted soil liner (Figure 2(b)); and (iii) landfills with a Subtitle D composite liner consisting of a geomembrane upper component and a compacted soil lower component (with the soil lower component having a hydraulic conductivity not greater than 1×10^{-7} cm/s) (Figure 2(c)). These minimum requirements indicate that less protective final cover systems are allowed at landfills with less protective liner systems, a logic questioned by many engineers. In practice, more protective cover systems are often used at unlined landfills, notwithstanding the leniency of the regulations. It should also be noted that the final cover systems described in the regulations do not represent "complete" designs. For example, the cover system shown in Figure 2(c) does not include a drainage layer above the geomembrane barrier layer, nor an adequate thickness of cover soil to allow sufficient water storage for vegetation. Good current practice would include both of these items.

Cover System Materials: Materials commonly used in landfill final cover systems are listed in Table 1. Analysis and design methods have been developed to evaluate the degree to which these materials, alone or in combination, can satisfy the functional requirements. Methods have also been developed to evaluate collateral issues associated with use of a particular final cover system in a given application, such as the stability of the system on slopes, integrity of the system under the applied stresses, and durability of the system. The design life of a cover system is usually assumed to be the post-closure care period, which for MSW landfills in the United States is defined by regulation to be at least 30 years, unless it can be demonstrated to the satisfaction of the appropriate regulatory agency that a shorter design life would be sufficient to protect human health and the environment. It is noted that issues associated with the durability characteristics of materials used in final cover systems are not addressed in detail in this paper.

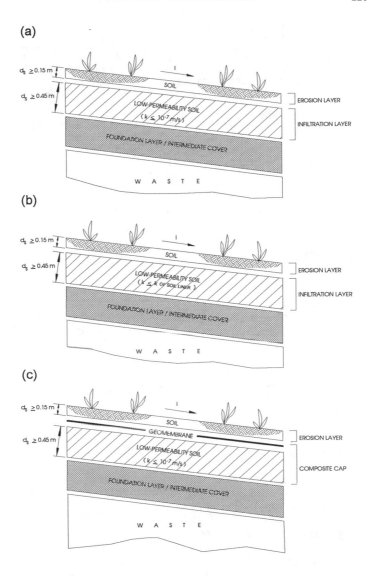

Figure 2.   USEPA (1992) recommended minimum final cover systems for: (a) unlined MSW landfills; (b) MSW landfills underlain by a soil liner; and (c) MSW landfills underlain by a geomembrane/soil composite liner. Note: d_s = soil layer thickness.

Table 1. Potential cover system materials[1].

• *Surface Layer*
  · Top Soil
  · Geosynthetic Erosion Control Layer
    Over Top Soil
  · Cobbles
  · Paving material
  · Others
 	
• *Protection Layer*
  · Soil
  · Cobbles
  · Others
 	
• *Drainage Layer*
  · Sand
  · Gravel
  · Geonet
  · Others
	• *Barrier Layer*
	  · Compacted Clay
	  · Geomembrane
	  · Geosynthetic Clay Liner (GCL)
	  · Geomembrane/Compacted Clay
	    Composite
	  · Geomembrane/GCL Composite
	  · GCL/Compacted Clay Composite
	  · Others
	
	• *Gas Collection Layer*
	  · Sand
	  · Gravel
	  · Geotextile
	  · Geonet
	  · Others
	
	• *Foundation Layer*
	  · Soil
	  · Select Waste
	  · Others

Notes: (1)   A soil or geotextile filter may be required between adjacent materials
to minimize migration of fines from one material to another.

Scope of Paper: The main categories of analysis and design related to MSW
landfill final cover systems are listed below. A discussion of each category is
presented in the remainder of this paper.

- hydraulic analysis;

- settlement analysis;

- static slope stability analysis;

- dynamic slope stability analysis; and

- surface-water management.

Hydraulic Analysis

Overview: A primary function of the final cover system is to limit the amount
of precipitation that infiltrates into the landfill. For MSW landfills, the amount of
leachate generated by a landfill is proportional to the amount of infiltration. The

drainage and barrier components of the cover system are used to limit infiltration. The functions of the barrier and drainage layers are complementary. The barrier layer impedes the migration of rainwater through the final cover and improves the performance of any overlying drainage layer. The drainage layer limits the buildup of hydraulic head on the underlying barrier layer and conveys collected water to drainage swales and/or perimeter drainage ditches. Hydraulic analyses are carried out to estimate infiltration rates. These analyses are typically performed using hydrologic computer models, though closed-form analytical solutions to estimate infiltration are also available. The analytical solutions may be used as a check of hydrologic model analysis results or for detailed design using conservative assumptions. These computer models and analytical solutions are described below

Hydrologic Computer Models: Several computer models are available to perform hydraulic analyses of landfill final cover systems. Three such models are: (i) Hydrologic Evaluation of Landfill Performance (HELP) model; (ii) Groundwater Loading Effects from Agricultural Management Systems (GLEAMS) model; and (iii) Water Balance Analysis Program (MBALANCE) model. The most widely used model is HELP. The descriptions of the models presented below are taken, in part, from Lane et al. (1992).

The HELP model simulates hydrologic processes for a landfill by performing daily, sequential water budget analyses using a quasi-two-dimensional, deterministic approach (Schroeder et al., 1994a,b). The hydrologic factors considered in the model include precipitation, surface-water storage (i.e., storage as snow), interception, surface evaporation, runoff, snow melting, infiltration, vegetation quality, evaporative zone depth, plant transpiration, soil evaporation, temperature, solar radiation, soil water storage, unsaturated flow, saturated flow, vertical drainage, lateral drainage, and vertical percolation through barrier layers. The HELP model uses five main routines to estimate runoff, evapotranspiration, vertical drainage to barrier layers, vertical percolation through barrier layers, and lateral drainage. Several other routines interact with the main routines to generate daily precipitation, temperature, and solar radiation values and simulate snow accumulation and melting, vegetative growth, interception, and percolation through geomembrane liners.

In the HELP model, runoff is computed using the runoff curve-number method of the U.S. Department of Agriculture Soil Conservation Service (USDA-SCS). Evapotranspiration is computed by a modified Penman method developed by Ritchie (1972) for initially wetted soil. Growth and decay of surface vegetation are modeled using a routine taken from the Simulator for Water Resources in Rural Basins (SWRRB) model (Arnold et al., 1989). Vertical drainage is computed using Darcy's equation, modified for unsaturated conditions using an unsaturated soil hydraulic conductivity calculated using an equation by Campbell (1974). Percolation through the soil barrier layer is also evaluated using Darcy's equation, but under saturated conditions. Infiltration through geomembranes and geomembrane/soil composite barriers is evaluated based on the work of Giroud and Bonaparte (1989a,b). A USEPA document entitled *"Design and Construction of RCRA/CERCLA Final Covers"* (USEPA, 1991) provides guidance on the use of the HELP model to evaluate landfill hydrologic performance.

The GLEAMS model is an updated version of the Chemical Runoff and Erosion from Agricultural Management Systems (CREAMS) model (Davis et al., 1990). This model was developed to simulate the effects of management systems on nonpoint source water pollution. The model predicts the water balance (e.g., runoff, evapotranspiration, and vertical drainage), sediment movement, and mass balance of chemicals (i.e., pesticides and nutrients) for an agricultural field. A field is defined as a management unit having: (i) a single land use; (ii) relatively homogeneous soil; (iii) spatially uniform rainfall; and (iv) a single management system (USDA-SCS, 1985). Normally, a field is less than 40 ha. The GLEAMS model is divided into three components, hydrology, erosion, and chemicals, which operate independently. Only the hydrologic component of this model will be discussed further.

The HELP model was partially adapted from CREAMS and, therefore, the theoretical bases for HELP and GLEAMS are similar. GLEAMS predicts the daily water balance considering runoff and runon calculated by the SCS curve-number procedure, evapotranspiration estimated using the modified Penman equation by Ritchie (1972), and vertical drainage within the root zone calculated using a soil storage routing technique described by Williams and Hann (1978). Like HELP, GLEAMS also considers snow accumulation and melting and vegetative growth.

The MBALANCE model was developed by the Wisconsin Department of Natural Resources to calculate the water balance of landfill covers (Kmet, 1982; Scharch, 1985). The model incorporates the Thornthwaite and Mather (1957) water balance procedure as adapted by Fenn et al. (1975). The MBALANCE program computes a monthly water balance by calculating surface runoff, evapotranspiration, and soil moisture change and subtracting these amounts from precipitation to yield percolation through the cover system.

A number of researchers have performed field studies and analytical assessments to evaluate the HELP, GLEAMS, and MBALANCE models (EPRI, 1984; Peters et al., 1986; Peyton and Schroeder, 1988; Barnes and Rodgers, 1988; Udoh, 1991; Lane et al., 1992; Benson et al., 1993; Peyton and Schroeder, 1993; Field and Nangunoori, 1994; Khire et al., 1994). In particular, the studies evaluated the reliability of the hydrologic models as tools to predict trends and magnitudes of the different landfill water balance components (i.e., infiltration, runoff, etc.). The conclusions of these studies are not in general agreement. For example, some of these studies found that HELP overpredicted infiltration in humid climates and underpredicted infiltration in arid climates, but other studies concluded just the opposite. In many cases, the models were not able to predict short-term trends.

However, for a number of cases they gave reasonable predictions of cumulative water balances.

In many of the comparisons between measured and calculated water balances, site-specific field data is used in the water balance predictions. However, in the current state of practice, measurement of site-specific parameters required for the models, such as soil field capacity, wilting point, and evaporation depth or rooting depth, is not performed. Thus, the model user is left to depend on default data which may lead to an inaccurate representation of a site. At present, these hydrologic models should be used carefully to develop conservative and reasonable bases for design. As a true predictive tool, the value of the models is limited unless site-specific calibrations are performed.

Analytical Solutions: Infiltration through the cover system may be calculated analytically using a two-step process: (i) estimate or calculate the average head of water in the drainage layer above the barrier layer; and (ii) calculate percolation through the barrier layer under the average head. Giroud et al. (1992) developed a differential equation to calculate the head of liquid in a drainage layer as a function of distance along the layer. A summary of this work has recently been published by Giroud and Houlihan (1995). They considered a drainage layer with a hydraulic conductivity, k, and horizontal length, L, on a slope forming an angle β with the horizontal (Figure 3). To derive the equation, the following assumptions were made: (i) the drainage layer was underlain by an impermeable barrier; (ii) precipitation, p, enters the layer at a constant, uniform rate (often taken as a percentage of the average annual rainfall); (iii) flow occurs under steady-state conditions; and (iv) the flow velocity is constant and small and, therefore, the flow is laminar and Darcy's equation can be applied. This equation must be solved numerically. Giroud and Houlihan provided tabulated results. For values of (p/k) < 0.25 $\tan^2\beta$, a condition that will be satisfied by most cover systems incorporating drainage and barrier layers, the average liquid thickness, T_{avg}, can be approximated as:

$$T_{avg} = \frac{p\,L}{2\,k\,\sin\beta} \qquad (1)$$

Basic SI units are: T_{avg} (m), p (m/s), L (m), k (m/s), and β (degree). The average hydraulic head, $h_{avg} = T_{avg}\cos\beta$.

Giroud et al. (1992) showed that Equation 1 provides better agreement with the numerical solution of the differential equation for head than does an equation in wide use today in U.S. landfill engineering practice (Moore, 1983). They also noted excellent agreement between their solution and results obtained using the HELP model.

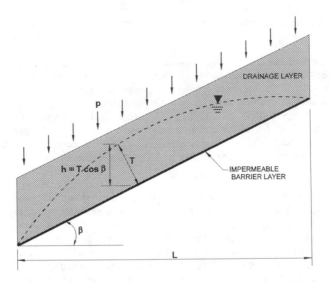

Figure 3.   Liquid thickness and hydraulic head in cover system drainage layer.

Steady-state percolation of rain water through a soil barrier layer can be calculated using Darcy's equation. For a geomembrane cap or a composite cap consisting of a geomembrane overlaying a soil barrier layer, steady-state percolation through the cap occurs essentially by infiltration through geomembrane holes. Equations to calculate infiltration through geomembrane and composite caps have been presented by Bonaparte et al. (1989), Giroud and Bonaparte (1989a,b), and Giroud et al. (1989, 1994). As previously noted, forms of these equations have been incorporated into the most recent version of the HELP model.

Settlement Analysis

Mechanisms of Settlement: Landfill final cover systems may be subject to settlements resulting from a variety of mechanisms. For purposes of evaluating cover system performance, settlements can be considered to have one of three sources, as illustrated in Figure 4. These sources are: (i) settlement of foundation soils; (ii) settlement due to overall waste mass compressibility; and (iii) settlement due to localized mechanisms. It is important to be able to estimate the magnitude and distribution of settlement from each source and the effects of settlement on the performance of final cover system components. Excessive angular distortion or differential settlement may: (i) induce unacceptable tensile stress and strain in final cover system components which may lead to component tearing or cracking; or (ii) cause final cover system slopes to change or reverse, which, in turn, may affect the performance of the final cover system drainage layer and/or the gas collection layer.

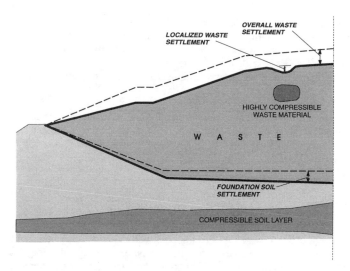

Figure 4.   Sources of landfill cover system settlements.

Settlement of Foundation Soils: Impacts of foundation settlement on the performance of the final cover system are usually insignificant. Occasionally, situations arise where foundation settlements are of sufficient magnitude to be considered in the design of a final cover system. For example, if the landfill is underlain by a thick layer of soft clay, consolidation settlements may be large. In these cases, design analyses should consider settlement of the clay layer. Both primary settlement and long-term secondary settlements should be considered. Calculations are performed using conventional equations from soil mechanics theory and a timeframe at least equal to the active life and post-closure care period of the unit.

Overall Waste Compression: Overall waste mass compressibility results in area-wide landfill settlements. Waste compression results from complex factors including (Sowers, 1973; Edil et al., 1990): (i) mechanical compression due to self-weight and surface loads; (ii) ravelling (i.e., movement of fines into larger voids); (iii) physico-chemical changes including corrosion, oxidation, and combustion; and (iv) bio-chemical decomposition under aerobic and anaerobic processes. The magnitude and rate of settlement are controlled by many factors among which are the waste fill height, organic content, age, moisture content, degree of compaction, and temperature.

A waste layer placed in a landfill cell experiences a relatively rapid mechanical settlement (sometimes referred to as "primary" settlement) under its own weight and the weight of layers above it.   This settlement is due to the compression, rearrangement, and ravelling of waste particles and usually occurs rapidly, within a few days to a few months (Sowers, 1973; NAVFAC, 1983; Burlingame, 1985). In addition to short-term mechanical settlement, waste settles as a result of long-term creep, physico-chemical reactions, and, most importantly for MSW, biodegradation. This time-dependent settlement is often referred to as "secondary" settlement. As with soils, the rate of secondary settlement is usually assumed to be independent of load.

A number of methods have been proposed for evaluating the time-independent and time-dependent compression of waste. Three methods that have been adopted from the soil mechanics literature are: (i) the one-dimensional compression model; (ii) the power creep model; and (iii) the Gibson and Lo model. These models are described below.

One-dimensional compression models are widely used to estimate waste settlements (Sowers, 1973; Burlingame, 1985; Landva and Clark, 1990; Fassett et al., 1994; Stulgis et al., 1995). According to these models, the total settlement of an element of solid waste due to an applied vertical stress may be taken as the sum of the primary and secondary settlement. One example of this type of model was presented by Sowers (1973). With this approach, primary settlement (ΔH_p) is calculated using an equation with the form:

$$\Delta H_p = C_{c\epsilon} H \log \left(1 + \frac{\Delta\sigma}{\sigma_o}\right) \tag{2}$$

where: $C_{c\epsilon}$ = modified primary compression index; H = initial height of waste; $\Delta\sigma$ = additional effective vertical stress applied to waste; and σ_o = initial effective overburden stress on waste. Basic SI units are: ΔH_p (m), H (m), $\Delta\sigma$ (kPa), σ_o (kPa), and $C_{c\epsilon}$ is dimensionless. It is noted that the effective stress terms are generally equal to corresponding total stresses because landfill waste materials are typically unsaturated.

The secondary waste settlement (ΔH_s) that occurs between times t_1 and t_2 is often calculated using an equation of the form:

$$\Delta H_s = C_{\alpha\epsilon} H_1 \log \frac{t_2}{t_1} \tag{3}$$

where: $C_{\alpha\epsilon}$ = modified secondary compression index; H_1 = height of waste at time t_1; t_1 = elapsed time from initial waste placement in the landfill to placement of final cover system; and t_2 = elapsed time from final cover system placement. Basic SI units are: ΔH_s (m), H_1 (m), t_1 (s), t_2 (s), and $C_{\alpha\epsilon}$ is dimensionless.

Values of the modified primary compression index and the modified secondary compression index for MSW have been reported to range from 0.05 to 0.4 and 0.01 to 0.1, respectively (Sowers, 1973; NAVFAC, 1983; Burlingame, 1985; Landva and Clark, 1990; Fassett et al., 1994; Stulgis et al., 1995). It should be noted that these values of compression indices were obtained from settlement data for older landfills where the waste may not have been very well compacted and where the potential for waste settlement may be large in comparison to well compacted waste placed in modern landfills.

Edil et al. (1990) proposed a power creep model for estimating landfill settlements. The form of the power creep equation is as follows:

$$\Delta H_t = H \, \Delta \sigma \; m \; (t/t_r)^n \tag{4}$$

where: ΔH_t = total waste settlement; $\Delta \sigma$ = average applied stress in the waste; m = reference compressibility; t = time since load application; t_r = reference time introduced into the equation to make time dimensionless; and n = rate of compression. Basic SI units are: ΔH_t (m), $\Delta \sigma$ (kPa), t (s), t_r (s), m (kPa^{-1}), and n is dimensionless. Edil et al. utilized settlement data from four different MSW landfills to back-calculate values for the two empirical parameters in Equation 4, m and n. Assuming a reference time, t_r, of 1 year, they calculated an average reference compressibility, m, of 2.5 x 10^{-5} kPa^{-1} and an average rate of compression, n, of 0.65.

Edil et al. (1990) also proposed use of the rheological model of Gibson and Lo (1961) for estimating landfill settlements. Using this approach, total waste settlement is expressed as:

$$\Delta H_t = H \, \Delta \sigma \; [a + b(1 - e^{-(\lambda/b)t})] \tag{5}$$

where: a = primary compressibility parameter; b = secondary compressibility parameter; and λ/b = rate of secondary compression. Basic SI units are: a (kPa^{-1}), b (kPa^{-1}), and λ/b (s^{-1}). Edil et al. fitted the Gibson and Lo model to settlement data for the same four MSW landfills used to calibrate the power creep equation parameters. They found the calibrated parameters a and b to decrease with increasing stress and λ/b to increase with increasing rates of waste compression. Edil et al. found the power creep equation to better fit the MSW settlement data than the Gibson and Lo model.

The process of waste settlement is highly complex due to the heterogeneous nature of the landfill and the erratic and continuous changes in the material properties and characteristics with time. The models described above represent relatively simple empirical mathematical formulations for describing the magnitude and rate of waste settlement. These models do not separate the components of waste settlement caused by the different settlement mechanisms (i.e., mechanical compression, ravelling, physico-chemical changes, and bio-chemical decomposition). Instead, the models combine the different components of waste settlement into one or two components that incorporate general compressibility parameters which are empirically determined.

The models, particularly the first one described above, are commonly used to predict waste settlement and, when properly applied, give reasonable estimates of ultimate settlements. More recently, attempts to model the effects of waste biodegradation on settlement have been made (e.g., Wall and Zeiss, 1995). Waste biodegradation models show promise for future use. However, to calculate total settlement, these methods must be calibrated and the contributions to total settlement resulting from other factors must be taken into account.

Settlement Due to Localized Mechanisms: Local depressions sometimes develop in landfill final cover systems. Landfill operators report that these depressions typically develop within the first several years after cover system installation. Such depressions are typically repaired (filled in) as a routine maintenance activity. In the extreme case, localized differential settlements could lead to excessive stresses in cover system components, although the authors are not aware of any reported cases. These depressions are generally attributed to several mechanisms, namely: (i) deterioration and collapse of objects in the waste just below the cover system (collapse mechanism); (ii) settlement associated with a highly compressible zone of waste (compression mechanism); and (iii) migration of fines from cover system soils into voids within the waste (migration mechanism).

Localized settlement mechanisms are usually addressed by appropriate landfilling practices and cover system design and construction measures. Appropriate landfilling practices include waste spreading and compaction, special attention to placement and compaction of collapsible objects, and placement of daily cover soil. Appropriate final cover system design and construction practices include:

• using soil and/or a select waste (i.e., waste free of collapsible objects) as a foundation layer for the final cover system to buffer the cover from the effects of underlying differential settlement;

• performing heavy proofrolling of the waste prior to constructing the final cover system;

• specifying final cover system slope inclinations large enough to accommodate differential settlements without grade reversal; and

• in special cases, installing a high-tensile-modulus geosynthetic reinforcement layer beneath the final cover system to mitigate adverse of differential settlement.

Typically, analyses to evaluate impacts of localized settlements on the final cover system are not performed as part of the final cover design. However, the authors have found that in certain cases it is necessary to perform design analyses to evaluate potential effects of localized areas of high waste compressibility on cover system performance. Situations where analyses may be called for include: (i) cover systems for abandoned, inactive landfills where the composition of waste is unknown or there is reason to believe that significant local waste heterogeneity may exist; and (ii) in response to requests from regulatory agencies for such evaluations. Several authors have described analysis methods that may potentially be used to evaluate localized waste settlements and their effects on final cover system performance. None of these methods have been field calibrated to any significant degree and selection of input parameters to the analyses is based primarily on engineering judgement.

Examples of methods that may be applied to analysis of localized waste settlement are as follows:

• Giroud et al. (1990) presented an approach based on soil arching theory for analyzing the stresses and strains in geosynthetics (such as geosynthetic layers within a cover system) that lose foundation support after construction due to development of a foundation void or depression. Poorooshasb (1991) used somewhat different analytical approach to address a similar problem. These methods may be applicable to the collapse and migration mechanisms described above.

• Jang and Montero (1993) used a boundary element formulation to model deformations around a collapsing void within an existing waste landfill, and to design the liner system of a planned vertical expansion of the landfill. This method may be applicable to the collapse mechanism.

• Carey et al. (1993) performed two-dimensional finite element analyses to evaluate the response of a landfill containing compressible zones and subjected to the loads imposed by a vertical expansion. Sagaseta (1987) presented a method for evaluating displacements in a uniform soil deposit subjected to a loss of soil mass near the ground surface. These methods may be applicable to the compression mechanism.

The authors of this paper have applied the Sagaseta method to evaluate displacements around a compressible zone in waste. Displacements and strains were calculated for a proposed cover system for a landfill with potential localized waste heterogeneities. Assumptions were made regarding the size of the compressible zone and its depth below the cover system, based on information regarding landfill operations and cover system design and construction procedures. Figure 5 presents results of this analysis. For this particular problem, grade reversal and tensile strains on the order of 0.8 percent were predicted in the geomembrane component of the cover system.

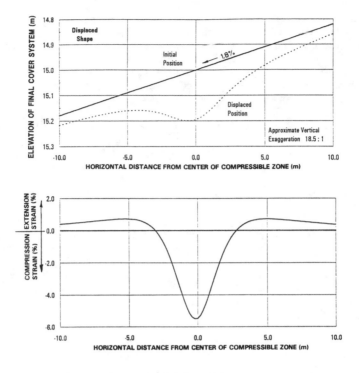

Figure 5.   Results of analysis of differential cover system settlements using the displacement method of analysis.

Impacts of Settlement on Cover System: In design, settlement profiles accounting for the various settlement mechanisms are developed to evaluate potential impacts to the final cover system. The evaluation usually considers: (i) post-settlement final cover grades, (ii) the potential for depressions in the final cover system, and (iii) stresses and strains in final cover system components. For a given deformed geometry, geosynthetic stresses and/or strains can be estimated using equations presented by Giroud (1984), Giroud et al. (1990), and Koerner and Hwu (1991).

Tensile stresses and strains induced in final cover system components should be less than the allowable tensile stresses and strains for these materials. Clays, geomembranes, and GCLs are typically used as barrier layers in final cover systems, as indicated in Table 1. The allowable tensile stresses and strains of these materials have been the subject of study by a number of researchers as described below.

Tensile strains causing cracking in compacted clays have been evaluated by Leonards and Narain (1963), Ajaz and Parry (1975a,b, 1976), Gaind and Char (1983), Chandhari and Char (1985), and LaGatta (1992). Based on these studies, compacted clays tested under unconfined conditions exhibit brittle behavior and reached failure at axial extensional strains of 0.02 to 4.4 percent with most clays exhibiting failure under unconfined conditions at extensional strains of 0.5 percent or less. The studies also showed that magnitude of tensile strain needed to cause cracking of compacted clays increases with increasing percentage of fines and water content.

The tensile behavior of geomembranes varies depending on the geomembrane material. Based on the work of Giroud (1984), the present state-of-practice in the U.S. for the design of polyethylene geomembranes is to limit the allowable geomembrane tensile strain to the short-term yield strain divided by a factor of safety. Factors of safety of 1.5 to 2 are typical, resulting in allowable tensile strains for design of high density polyethylene (HDPE) geomembranes in the range of approximately 5 to 10 percent. More recently, Berg and Bonaparte (1993) presented a method for calculating the long-term allowable tensile stresses for polyethylene geomembranes which accounts for time, temperature, exposure environment, seam response, and boundary-stress and deformation conditions.

LaGatta (1992) and Boardman (1993) evaluated the impact of differential settlement on the hydraulic conductivity of geosynthetic clay liners (GCLs). The results of these evaluations indicate that GCLs can withstand substantial angular distortion and tensile strains of up to 10 to 15 percent without significant increase in hydraulic conductivity.

Static Slope Stability Analysis

Methods of Analysis: As part of landfill cover system design, analyses are typically carried out to evaluate the static stability of the final cover system. The considered failure mechanism is downslope sliding of one or more cover system components due to inadequate component shear strength or inadequate shear strength at the interface between adjacent components. In most cases, the potential sliding mass can be represented as two-dimensional infinite slope or two-part wedge. The limit equilibrium (LE) analysis method is widely used to evaluate static slope stability. Another analysis method, the force equilibrium and strain compatibility (FESC) method, is used occasionally. Both of these methods are described below.

When applying the LE method to cover systems, the stability of a soil mass above a potential sliding surface is evaluated by calculating a slope stability factor of safety (i.e., the ratio of the available shear resistance to the applied shear stress). Several LE formulations are available to evaluate the stability of soil-geosynthetic systems on slopes. The simplest formulation is to assume infinite slope conditions and purely frictional material and interface shear strengths. This approach ignores the stabilizing influences of passive soil resisting forces at the toe of the slope and any apparent cohesion/adhesion in cover system materials and interfaces. Furthermore, this approach does not account for tension in the geosynthetic layers.

For many cover system geometries, infinite slope equations provide a reasonable, simple, and slightly conservative basis for design. The slope stability factor of safety, considering seepage forces parallel to the slope above the potential sliding surface, is:

$$FS = (1 - \frac{T_w}{T_c} \frac{\gamma_w}{\gamma}) \frac{\tan\phi_i}{\tan\beta} \tag{6}$$

where: FS = factor of safety; T_w = thickness of water above the sliding surface; T_c = thickness of soil cover above the sliding surface; γ_w = unit weight of water; γ = wet unit weight of soil (assumed equal to the saturated unit weight of the soil); ϕ_i = friction angle of interface (i.e., along the sliding surface); and β = slope angle. Basic SI units are: T_w (m), T_c (m), γ_w (kN/m^3), γ (kN/m^3), ϕ_i (degree), and β (degree). These dimensions and parameters are illustrated in Figure 6.

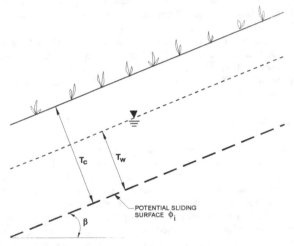

Figure 6.   Dimensions and parameters for cover system infinite slope stability analysis.

Seepage forces should be included in the design calculations if the cover system does not include an effective drainage layer above the barrier layer. However, the current state of practice is to include a drainage layer and this practice is strongly encouraged. For a cover system which incorporates an effective drainage layer or for analysis of sliding surfaces located below a geomembrane in the cover system, effect of seepage forces resulting from water flow above the geomembrane on stability is negligible. In this case, the factor of safety (FS) can be approximated as $\tan\phi_i$ / $\tan\beta$.

Giroud et al. (1995) presented an equation to calculate the factor of safety of a thin soil veneer on a slope assuming that the resistance to sliding is provided by mobilized shear resistances at the toe (i.e, buttress resistance) and along the potential sliding surface and by a mobilized tensile force in any geosynthetic materials that are located above the potential sliding surface and anchored outside the sliding mass (Figure 7(a)). Giroud et al. modeled the potential sliding mass using a two-part wedge and assumed that the shear resistance along the entire sliding surface is represented by both friction and cohesion terms, the sliding mass has constant thickness, and no seepage forces or pore pressures are present. Using these assumptions, they developed the following equation:

$$FS = \frac{\tan\phi_i}{\tan\beta} + \frac{c_i}{\gamma\,T_c\,\sin\beta} + \frac{T_c}{H}\frac{\sin\phi_c}{\sin(2\beta)\,\cos(\beta+\phi_c)}$$

$$+ \frac{C_c}{\gamma\,H}\frac{\cos\phi_c}{\sin\beta\,\cos(\beta+\phi_c)} + \frac{\alpha}{\gamma\,H\,T_c} \tag{6}$$

where: H = height of soil cover; ϕ_c = friction angle of cover soil; c_c = cohesion of cover soil; c_i = cohesion of interface; and α = tension in geosynthetics. Basic SI units are: H (m), ϕ_c (degree), c_c (kPa), c_i(kPa), and α (kN/m). These dimensions and parameters are illustrated in Figure 7(b).

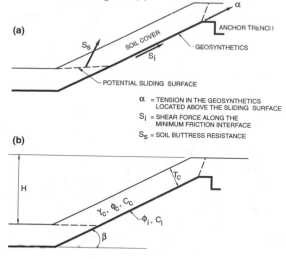

Figure 7.   Two-part wedge stability analysis of final cover system: (a) forces contributing to stability; and (b) relevant dimensions and parameters.

Several other closed-form applications of the LE method to layered soil-geosynthetic systems on slopes have been proposed (e.g., Koerner and Hwu, 1991; Bourdeau et al., 1993; Druschel and Underwood, 1993). In addition, most commercially available slope stability computer software programs (e.g., STABL (Siegel, 1975), UTEXAS3 (Wright, 1991), XSTABL (Sharma, 1994)) also incorporate one or more general formulations of the LE method that are used to perform static analysis of final cover systems. The general LE formulations in the available software have the advantage that they can be applied to non-uniform slope and soil cover conditions.

The LE method is useful for evaluating cover system stability under most conditions but is subject to several limitations. With the LE method, material and interface shearing resistances are assumed to be independent of displacement. For many materials and interfaces, however, mobilized shearing resistance increases with increasing displacement to a peak value. As the displacement increases beyond that required to mobilize peak strength, the shear resistance decreases to a large displacement and ultimately "residual" value (i.e., "strain-softening"). In using the LE method, judgement must be applied to the selection of shear strength values for strain-softening materials (i.e., peak, residual, or some intermediate value). The LE method is similarly limited with respect to tension forces in cover system geosynthetic components, and therefore, cannot be used to estimate the magnitude and distribution of stresses and strains in those components.

FESC methods are sometimes used for cover system design when the limitations of LE methods may be significant. The primary advantage of FESC methods is their ability to account for the stress-strain response of materials and interfaces and, therefore, to predict the distribution of stresses and strains within the cover system components, particularly geosynthetic components. FESC methods can also account for the effects of construction sequencing. The primary disadvantage of FESC methods is the relatively large effort required to obtain stress-strain relationships and perform the calculations.

Several published studies are available on the application of FESC methods to cover systems. For example, Long et al. (1993, 1994) describe a finite difference model (GEOSTRES) that considers force equilibrium and strain compatibility. GEOSTRES uses inelastic, non-linear springs to model the shear resistance-displacement behavior at each interface and to model the axial load-displacement behavior within each component. Another example of a FESC method is given by Wilson-Fahmy and Koerner (1992, 1993). They adopted a two-dimensional finite element approach to account for force equilibrium and strain compatibility in stability analysis of soil-geosynthetic systems on slopes.

Static Shear Strength Parameters: For conceptual design, presumptive static shear strength parameters for final cover system components and interfaces are often used based on information available in the literature or based on experience at other sites.

A number of publications are available which include results of laboratory shear strength tests on soil-geosynthetic and geosynthetic-geosynthetic interfaces. These include Martin et al. (1984), Williams and Houlihan (1987), Long et al. (1993), and Nataraj et al. (1995). Prior to final design, laboratory testing using site-specific materials and testing procedures and conditions is typically performed to establish design shear strength parameters for cover system materials and interfaces. The authors highly recommend project specific interface shear testing for detailed design. The use of presumptive shear strengths for final design should be avoided.

Various methods for testing shear strength of soils in the laboratory and in the field are well-known and are fully described in a number of textbooks and laboratory guides. There are several devices currently used to test shear strength of soil-geosynthetic system components and interfaces including: (i) the large-scale direct shear box; (ii) the conventional direct shear box; (iii) the torsional shear device; and (iv) the tilt table. Gilbert et al. (1995) provided an overview of these devices. Table 2 (from Gilbert et al., 1995) summarizes the advantages and disadvantages of the different devices.

Table 2. Summary of Advantages and Disadvantages Associated with Test Devices for Measuring Interface Shear Strength (Gilbert et al., 1995).

Test Device	Advantages	Disadvantages
Large-Scale Direct Shear Box	• Industry Standard • Large Scale • Large Displacements • Minimal Boundary Effects	• Machine Friction • Load Eccentricity • Limited Continuous Displacement • Limited Normal Stresses • Expensive
Conventional Direct Shear Box	• Experience with Soil • Inexpensive	• Machine Friction • Load Eccentricity • Small Scale • Limited Displacements • Boundary Effects
Torsional Shear Device	• Unlimited Continuous Displacement	• Machine Friction • Anisotropic Shearing • Small Scale • Expensive
Tilt Table	• Minimal Machine Effects • Minimal Boundary Effects • Inexpensive	• Small Experience Base • Limited Continuous Displacement • Limited Normal Stresses • No Post-Peak Behavior

Project-specific shear strength testing programs are designed to simulate the anticipated field conditions by selecting appropriate values for testing procedures and conditions. These include the soil compaction conditions (i.e., water content and density), soil consolidation load and time, the wetting conditions for the materials and interfaces, the range of applied normal stresses, the direction of shear for geosynthetic interfaces, and the shear displacement rate and magnitude. The potential effects of many of these testing conditions on measured interface shear strength parameters are reported in the literature (e.g., Collios et al., 1980; Ingold, 1982; Myles, 1982; Saxena and Wong, 1984; Akber et al., 1985; Degoutte and Mathieu, 1986; Miyamori et al., 1986; Eigenbrod and Locker, 1987; Fourie and Fabian, 1987; Williams and Houlihan, 1987; Seed and Boulanger, 1991; Seed et al., 1988; Seed, 1989; Swan et al., 1991; Pasqualini et al., 1993; Stark and Poeppel, 1994; Bemben and Schulze, 1995).

Several additional factors may affect long-term shear strength properties of the cover system components and interfaces. For example, in cold regions, freeze-thaw may reduce the shear strength of cover system clays and clay/geosynthetic interfaces. Research has shown that many compacted clays undergo significant change in soil fabric and decrease in shear strength as a result of freeze-thaw cycling (e.g., Nagasawa and Umeda, 1985). In addition, both heating and cooling result in soil moisture migration which can cause changes in component and interface shear strengths (e.g., Daniel et al., 1993). Furthermore, long-term creep may also be significant, particularly in geosynthetic components. No consistent standard of practice presently exists for directly addressing the potential effects of these factors on cover system stability. These factors may be indirectly accounted for through use of higher minimum acceptable factors of safety, when appropriate, or through placement of an additional thickness of cover soil above the critical layers for thermal insulation and isolation from environmental factors.

Factor of Safety: LE analysis methods provide a calculated slope stability factor of safety. Minimum acceptable slope stability factors of safety for cover systems depend on project-specific conditions and uncertainties. For example, when cover systems include strain-softening components or interfaces, differing minimum factors of safety are often applied for peak strength analysis and analysis based on residual strength. In addition, several additional factors typically influence selection of the minimum acceptable factor of safety for static slope stability including regulatory requirements, reliability of laboratory test methods, similarity between laboratory testing conditions and field conditions, reliability of method of slope stability analysis used, construction procedures and quality control and assurance, and consequences of slope failure. General guidance on selecting factors of safety for slope stability analysis can be found in USACOE (1970), Mitchell and Mitchell (1992), and Duncan (1992). The authors recommend that in all cases involving strain-softening materials or interfaces, a check be made to assume a minimum factor of safety of at least 1.1 to 1.2 based on residual shearing conditions.

Dynamic Slope Stability Analysis

Overview: The final cover system of a landfill may be subject to damage as a result of strong ground accelerations that can accompany an earthquake. State and federal regulations require evaluation of the impacts of these accelerations on containment systems for landfills located in seismic impact zones. These regulations (40 CFR §258) define a seismic impact zone as an area having a 10 percent or greater probability that the peak horizontal acceleration in lithified earth material (i.e., bedrock), expressed as a percentage of the earth's gravitational pull (g), will exceed 0.10 g, within a 250-year period.

Due to the potential for amplification of peak horizontal accelerations between the ground surface at the base of the landfill and the top of the landfill, final cover systems can potentially be damaged by peak horizontal accelerations smaller than the 0.10 g regulatory value. For this reason, seismic impact analysis are occasionally performed for landfills that are not within regulatory seismic impact zones. This is most often the case for landfills with final cover systems containing components and interfaces with low shear strengths when placed on relatively steep landfill slopes.

Evaluation of dynamic stability of a landfill final cover system typically involves the following three steps: (i) estimating peak horizontal accelerations at the top of the landfill; (ii) selecting dynamic shear strength properties for cover system components and interfaces; and (iii) performing dynamic slope stability analysis. In a recent USEPA document entitled "RCRA Subtitle D (258) Seismic Design Guidance for Municipal Solid Waste Landfill Facilities" (USEPA, 1995), methods to calculate peak horizontal accelerations at a landfill and methods to evaluate the seismic slope stability of the landfill containment systems are presented. The methods and practices described below for landfill final cover systems are believed to be consistent with those presented in the USEPA document.

Peak Horizontal Acceleration: The peak horizontal acceleration at the top of the landfill is typically estimated using a United States Geologic Survey (USGS) seismic probability map and/or detailed one-dimensional site response analysis. The USGS seismic probability maps present peak bedrock accelerations and velocities having a 90 percent probability of not being exceeded over 10, 50, and 250 year periods (USGS, 1982, 1990). If bedrock is not present at or near the ground surface, the peak acceleration from the USGS maps may have to be modified for site specific conditions, as described below.

Several attempts have been made to relate the peak bedrock acceleration to the peak ground acceleration at a given site as a function of the local soil conditions (Seed and Idriss, 1982; Idriss, 1990; Kavazanjian and Matasović, 1995; Singh and Sun, 1995). A comparison of peak horizontal bedrock accelerations and peak horizontal accelerations at soft soil sites and landfill sites is shown in Figure 8 (based on Kavazanjian and Matasović, 1995). This comparison includes the relationship on Kavazanjian and Matasović, 1995). This comparison includes the relationship developed by Idriss (1990) and the upper-bound relationship prescribed by Harder (1991) for earth dams. Examination of Figure 8 indicates that the relationship of Idriss (1990) for soft soils provides a reasonable representation of average amplification potential of solid waste landfills. It is noted that the Harder (1991) upper-bound relationship for earth dams may also represent an upper-bound relationship for potential amplifications at solid waste landfills.

A procedure to perform a simplified site response analysis using the relationship of Idriss (1990) was described in the USEPA (1995) document. The purpose of this procedure is to adjust the peak acceleration from the USGS map for the influence of local soil conditions (to obtain the free field peak acceleration at the project site) and for propagation of accelerations upward from the bottom of the landfill (to obtain the peak acceleration at the crest of the landfill). This procedure is likely to be conservative for typical landfills in most areas of the United States.

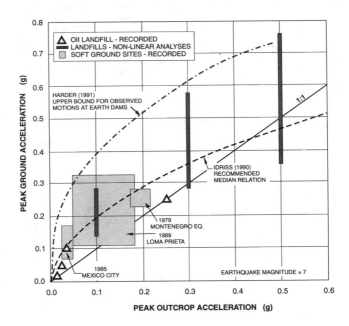

Figure 8.   Comparison of peak horizontal bedrock accelerations and peak horizontal accelerations at soft soil sites and landfill sites (based on Kavazanjian and Matasović, 1995).

A formal one-dimensional seismic site response analysis is typically performed in place of the simplified analysis for points where significant cover system accelerations are anticipated. Such projects may include those in regions where large earthquakes may occur, where soft foundation soil conditions are present, or where landfill waste thickness are relatively thin. The formal response analysis can be performed for the foundation soils only, for the waste mass only, or for both the foundation soils and waste mass, depending on project-specific conditions. It is noted that in rare cases two-dimensional site response analyses have been performed.

The computer program, SHAKE, originally developed by Seed and his coworkers (Schnabel et al., 1972) and recently updated by Idriss and Sun (1992) is perhaps the most commonly used computer program for one-dimensional seismic site response analysis. Basic input to SHAKE includes the soil profile, soil properties, and an earthquake acceleration time history. Soil properties include the maximum (small strain) shear wave velocity or shear modulus and unit weight for each soil layer plus curves relating the reduction in modulus and the fraction of critical damping to shear strain for each soil type. These soil properties are well-documented in the literature. Information is also available on the dynamic properties of MSW (e.g., Sharma et al., 1990; Fassett et al., 1994; Kavazanjian and Matasović, 1995; and Kavazanjian et al., 1995).

<u>Dynamic Shear Strength Parameters:</u> Estimates of the dynamic shear strength parameters of the components and interfaces of the landfill final cover system are needed to perform seismic slope stability analysis. These estimates are typically based on measured static undrained shear strength parameters. Shaking table laboratory test results and observed earthquake performance form the basis for such estimates.

The dynamic shear strength of clays is influenced by the amplitude of the seismic deviator stress, the number of applied cycles, and the plasticity of clay (Makdisi and Seed, 1978). In most cases (i.e., unsaturated compacted clay), the static shear strength is used as the shear strength for dynamic analyses. For saturated soft clays, the dynamic shear strength can conservatively be assumed to equal approximately 80 percent of the static undrained strength. However, for sensitive and stiff clays, cyclic loading can lead to reductions in strength and accumulated deformations can exceed the strain at which peak strength is mobilized, resulting in reduction of strength to residual values.

The cyclic shear strength of dry sand can be assumed to equal the static shear strength of the sand. When saturated, however, sand can undergo significant strength reduction. Evaluation of the potential for shear strength reduction in a saturated sand subject to dynamic loading may require sophisticated cyclic laboratory testing. Alternatively, a residual shear strength may be assigned to the sand based upon either undrained laboratory tests or in situ test results. Use of residual shear strengths in a dynamic stability assessment can result in a very conservative assessment of the pseudo-static factor of safety and/or yield acceleration.

The interface friction angle between geosynthetic components, (e.g., geomembrane and geotextile) from static laboratory tests is typically used for dynamic analyses. Some investigations have reported slight differences between static and dynamic interface strengths (Kavazanjian et al., 1991). Others have reported that interface shear strengths appear to be independent of the frequency content and number of cycles of motion (Yegian and Lahlaf, 1992). Augello et al. (1995) performed back analyses of landfill performance during the 17 January 1995 Northridge earthquake. The results of their analyses suggest that static shear strength measured in laboratory direct shear tests may be conservatively used to represent the dynamic interface shear strength. For interfaces that exhibit strain-softening in static shear strength tests, it may be advisable to consider residual strengths for dynamic analyses although this may be excessively conservative if little movement occurs during shaking.

With respect to soil-geosynthetic interfaces, Yegian et al. (1995) performed shaking table tests to determine the dynamic shear strength of a smooth geomembrane/Ottawa sand interface. They found the dynamic shear strength of this interface to equal its static shear strength. It should be noted that almost no information is available in the literature on the dynamic shear strengths of compacted clay/geosynthetic layer interfaces. Typical practice is to assume that the relationship between static undrained strength and dynamic strength for clay/geosynthetic interfaces is similar to that of compacted clay.

Methods of Analysis: Most of the static LE slope stability analysis methods discussed in the preceding section of this paper are also suitable for seismic stability analyses (Duncan, 1992). They can be used in conjunction with several different approaches for seismic analysis, of which the following two represent the current state of practice: (i) the pseudo-static approach; and (ii) the permanent seismic deformation approach.

In the pseudo-static factor of safety approach, a seismic coefficient is included in the LE analysis to represent the effect of the horizontal acceleration imposed by the earthquake upon the potential failure mass. The main drawback of the pseudo-static factor of safety approach lies in the difficulty in relating the value of the seismic coefficient to the characteristics of the design earthquake. Use of the peak acceleration at the crest of the landfill (expressed as a fraction of gravity) as the seismic coefficient for cover system stability analysis typically provides excessively conservative assessments of slope stability. Another drawback of the approach is the difficulty in assessing the potential significance of calculated factors of safety less than one. In this situation, a permanent seismic deformation analysis is typically performed, as discussed in the following paragraph.

In the permanent seismic deformation approach cumulative seismic deformations are calculated on the basis of the acceleration time history of the cover system and the acceleration that results in a factor of safety of 1.0 in a pseudo-static LE analysis (i.e, yield acceleration). With this approach, the final cover system is often modeled as an infinite slope. For the infinite slope case the yield acceleration for the cover may be assessed using the following general equation for the stability of an infinite slope proposed by Matasović (1991):

$$k_y = \frac{\dfrac{c}{\gamma\ z\ \cos^2\beta} + \tan\phi\ [1 - \dfrac{\gamma_w\ (z-d_w)}{\gamma\ z}] - \tan\beta}{1 + \tan\ \beta\ \tan\phi} \qquad (8)$$

where: k_y = yield acceleration, γ = wet unit weight of soil cover, γ_w = unit weight of water, c = cohesion along assumed failure surface, ϕ = angle of internal friction along assumed failure surface, β = slope angle, z = depth to the assumed failure surface, and d_w = depth to water surface (assumed parallel to the slope). Basic SI units are: k_y (fraction of gravity), z (m), d_w (m), β (degree), ϕ (degree), γ (kN/m³), γ_w (kN/m³), and c (kPa).

Permanent seismic deformations are typically calculated using the approach suggested by Newmark (1965) wherein acceleration pulses in the acceleration time history that exceed the yield acceleration are double integrated to yield cumulative relative displacements. In a Newmark analysis, permanent displacement is assumed to accumulate in one direction only, the downslope direction. The yield acceleration in the other (upslope) direction is assumed to exceed the peak upslope acceleration of the failure mass.

In practice, both the pseudo-static factor of safety and permanent seismic deformation approaches can be combined in a united seismic slope stability and deformation analysis. The united analysis involves an initial psuedo-static analysis followed by a deformation analysis, if warranted. This united analysis is described in detail in USEPA (1995).

Prevailing practice for landfill design is that a calculated permanent seismic deformation less than 150 to 300 mm is considered acceptable for landfill liner systems (Seed and Bonaparte, 1992; Anderson and Kavazanjian, 1995). For landfill final cover systems, larger permanent seismic deformations may be acceptable because cover system deformations are often readily observed in post-earthquake inspections and damage to the final cover system components is repairable (USEPA, 1995). It is noted that Anderson and Kavazanjian (1995) suggest that calculated deformations as large as 300 mm may begin to approach the limits of applicability of the Newmark deformation method.

Surface-Water Management

Erosion and Sediment Control: The surface layer of the final cover system will be exposed to environmental conditions over the design life of the landfill. During this time period, erosion and sediment control measures are implemented to minimize soil loss due to rain, runoff, and wind, prevent the development of rills or gullies, protect the underlying layers of the cover system including the barrier layer, and minimize sediment impact to the surrounding environment.

According to the U. S. Department of Agriculture (USDA) universal soil loss equation, soil loss is related to rainfall intensity, soil type, length and inclination of slope, vegetation, and land management practices. To reduce soil loss, the following measures are typically implemented.

- Soils or any other materials used in the surface layer of the final cover system are selected to be of appropriate texture, organic content, salt content, acidity (i.e., pH), and nutrient content for plant growth depending on the plant species and climatic conditions at the site.

- The length and steepness of final cover system slopes are minimized to reduce runoff velocity and therefore reduce erosion. The effective length of slopes is also controlled by using benches.

- Diversion structures are used to prevent surface water from running onto the final cover from adjacent areas.

- Drainage courses and outlet structures are used to handle concentrated or increased runoff.

- Sediment basins and traps, straw bale dikes, silt fences, or other effective sediment control measures are used to control sediment migration off-site.

Furthermore, landfill operators often:

- time placement and grading of final cover to coincide with a period of lower erosion potential as related to seasonal variations in precipitation;

- minimize the size of areas and length of time the final cover will be unprotected;

- stabilize areas on which final cover has been placed using vegetation and mulching or any other effective method;

- repair rills and gullies if they develop; add cover material, regrade, mulch, and reseed as needed;

- place erosion and sediment control measures out of the way of construction equipment to prevent any damage to the measures;

- inspect and maintain erosion and sediment control measures, particularly during and after storms, to ensure that they function properly and to correct any problems that may arise as soon as they develop; and

- plan for emergencies by stocking necessary erosion and sediment control supplies on-site and by having necessary equipment available to quickly correct problems that may arise.

Plant selection is region and project specific, but will typically involve meeting the following requirements (Goldman et al., 1986).

- Plants should have dense growth and fibrous roots to provide a continuous cover with minimum exposure.

- Plants should be able to germinate and grow rapidly to a size and areal extent that can provide good erosion protection.

- Plants may be required to tolerate the effects of landfill gas, depending on the gas permeability of the cover system.

- Plants should be able to adapt to site climate. Plants used to vegetate the surface layer are typically grasses. Grasses that have been found to tolerate local climatic conditions (e.g., sunshine, exposure, temperature, rainfall, drought, and wind) and local soil types should be used. Information on suitable grasses for a site is often available from the local USDA-SCS or county engineer.

- Plants should not require excessive irrigation and fertilization or mowing.

- Plants should not produce excessive fuel volumes (i.e., produce dry vegetation that may be a fire hazard).

- A mix of plants may be used, if needed, to enhance the chances of survival and assure successful vegetation.

- Plants should be resistant to diseases, insects and other pests.

- No trees, woody shrubs, or deep rooted plants should be planted or allowed to grow on revegetated areas. Rooting depth of any vegetation growing on the cover should not exceed the depth to the material which functions as a barrier layer.

The surfaces of graded areas are typically roughened before seeding and soil is loosened to a depth of approximately 50 mm. Seeding of the surface layer is best performed when temperature, moisture, and sunlight are adequate. Seeds should be evenly distributed and protected from erosion until adequate germination occurs. Fertilizers may be used where necessary to ensure successful vegetation. A local expert can be consulted regarding fertilizer requirements and specifications.

Mulching is typically used to control erosion until plants are large enough to do the job, to hold fertilizer, seed, and topsoil in place in the presence of wind, rain, and runoff, to promote germination of seeds, to moderate temperatures, and to increase the moisture retention of the soil. Materials that could be used include, but are not limited to, straw, wood fiber, wood chips, bark, and jute, textile, or plastic erosion-control mats.

Surface Drainage System: Storm-water runoff from the final cover system should be controlled using a surface drainage system. The drainage system may consist of a network of terraces, ditches, downchutes, and culverts. Each component of the drainage system should be designed for peak flows anticipated from the design storm. The design storm is usually specified for temporary and permanent conditions in federal, state, and local landfill, flood control, and soil conservation regulations. Flows from the landfill are typically directed to sediment traps, basins and/or ponds to minimize the release of sediments and control rates of water flow off site.

Several urban drainage models are available which can be used for surface water analysis and design of landfill drainage systems. Two of the more commonly used models are: (i) the SCS Technical Release Number 55 (TR55) entitled "Urban Hydrology for Small Watersheds" (USDA-SCS, 1986); and (ii) the U.S. Army Corps of Engineers (USACOE) Hydrologic Engineering Center (HEC) computer program entitled "HEC-1 Flood Hydrograph Package" (USACOE, 1990).

Landfill surface drainage system design typically involves the following general steps: (i) the landfill surface is divided into several distinct drainage areas, as necessary; (ii) the hydrologic properties of each area are estimated including size, soil type, and vegetative cover type; (iii) the peak runoff from the design storm is evaluated for each drainage area and surface drainage system component; and (iv) the size of each component of the surface drainage system is selected to handle the estimated peak flows associated with it.

Summary

The current state of practice regarding the design of conventional final cover systems for MSW landfills in the United States was summarized in this paper. Brief descriptions of design methods and practices which are commonly used by the general engineering community were provided. Where applicable, the advantages and disadvantages of using more sophisticated methods were also discussed. The major design aspects which were considered in this paper relate to: (i) flow of water in and through the final cover system; (ii) impacts of waste settlement on the performance of final cover system components; (iii) static and dynamic cover system stability; and (iv) surface-water management.

Acknowledgements

The authors thank Drs. Neven Matasović, Edward Kavazanjian, and Jean-Pierre Giroud, all of GeoSyntec Consultants, for their valuable contributions to the paper.

References

Ajaz, A., and Parry, R.H.G. (1975a). "Stress-Strain Behavior of Two Compacted Clays in Tension and Compression." *Geotechnique*, 25(3), 495-512.

Ajaz, A., and Parry, R.H.G. (1975b). "Analysis of Bending Stresses in Soil Beams." *Geotechnique*, 25(3), 586-591.

Ajaz, A., and Parry, R.H.G. (1976). "Bending Test for Compacted Clays." *Journal of Geotechnical Engineering Division*, ASCE, 102(GT9), 929-943.

Akber, S.Z., Hammamji, Y., and Lafleur, J. (1985). "Frictional Characterization of Geomembranes, Geotextiles, and Geomembrane/Geotextile Composites." *Proceedings of the Second Canadian Symposium on Geotextiles and Geomembranes*, Edmonton, Alberta, Canada, 209-217.

Anderson, D.G., and Kavazanjian, E., Jr. (1995). "Performance of Landfills Under Seismic Loading." *Proceedings of the Third International Conference on Recent Advances in Geotechnical Earthquake Engineering and Soil Dynamics*, University of Missouri, Rolla, MO, 3, Apr.

Arnold, J.G., Williams, J.R., Nicks, A.D., and Sammons, N.B. (1989). "*SWRRB, a Simulator for Water Resources in Rural Basins.*" Agricultural Research Service, USDA, Texas A&M University Press, College Station, TX.

Augello, A.J., Matasović, N., Bray, J.D., Kavazanjian, E., Jr., and Seed, R.B. (1995). "Evaluation of Solid Waste Landfill Performance During the Northridge Earthquake." *Earthquake Design and Performance of Solid Waste Landfills*, ASCE Geotechnical Special Publication (in press).

Barnes, F.J., and Rodgers, J.E. (1988). "*Evaluation of Hydrologic Models in the Design of Stable Landfill Covers.*" U.S. Environmental Protection Agency, Office of Research and Development, Washington, D.C., Project Summary, Report No. EPA/600/S-88/048.

Bemben, S.M., and Schulze, D.A. (1995). "The Influence of Testing Procedures on Clay/Geomembrane Shear Strength Measurements." *Proceedings of the Geosynthetics '95 Conference*, Nashville, TN, Feb, 3, 1043-1056.

Benson, C.H., Khire, M.V., and Bosscher, P.J. (1993). "*Final Cover Hydrologic Evaluation, Final Cover-Phase II.*" Environmental Geotechnics Report No. 93-4, University of Wisconsin, Madison, WI, 151 p.

Berg, R.R., and Bonaparte, R. (1993). "Long-Term Allowable Tensile Stresses for Polyethylene Geomembranes." *Geotextiles and Geomembranes*, 12, 287-306.

Boardman, B.T. (1993). "*The Potential Use of Geosynthetic Clay Liners as Final Covers in Arid Regions.*" M.S. Thesis, University of Texas, Austin, TX.

Bonaparte, R., Giroud, J.P., and Gross, B.A. (1989). "Rates of Leakage Through Landfill Liners." *Proceedings of the Geosynthetics '89 Conference*, San Diego, CA, 1, 18-29.

Bourdeau, P.L., Ludlow, S.J., and Simpson, B.E. (1993). "Stability of Soil-Covered Geosynthetic-Lined Slopes: A Parametric Study." *Proceedings of the Geosynthetics '93 Conference*, Vancouver, Canada, Feb, 1511-1521.

Burlingame, M.J. (1985). "Construction of a Highway on a Sanitary Landfill and Its Long-Term Performance." *Transportation Research Record 1031*, TRB, Washington, D.C., 34-40.

Campbell, G.S. (1974). "A Simple Method for Determining Unsaturated Hydraulic Conductivity from Moisture Retention Data." *Soil Science*, 117(6), 311-314.

Carey, P.J., Koragappa, N., and Gurda, J.J. (1993). "Renewal of Old Sites - Overliner System Design, A Case Study of the Brookhaven Landfill, Long Island, New York." *Proceedings of the Waste Tech '93 Conference*, National Solid Waste Management Association, Marina Del Rey, Jan.

Chandhari, A.P., and Char, A.N.R. (1985). "Flexural Behavior of Reinforced Beams." *Journal of Geotechnical Engineering*, ASCE, 111(11), 1328-1333.

Collios, A., Delmas, P., Gourc, J.P., and Giroud, J.P. (1980). "Experiments on Soil Reinforcement With Geotextiles." *The Use of Geotextiles for Soil-Improvement*, 80-177, ASCE National Convention, Portland, OR, 53-73.

Daniel, D.E., Shan, H.-Y., and Anderson, J.D. (1993). "Effects of Partial Wetting on the Performance of the Bentonite Component of a Geosynthetic Clay Liner." *Proceedings of the Geosynthetics '93 Conference*, Vancouver, Canada, Feb, 1483-1496.

Davis, F.M., Leonard, R.A., and Knisel, W.G. (1990). "*GLEAMS User Manual-Version 1.8.55.*" USDA-ARS Southeast Watershed Research Laboratory and University of Georgia, Lab Note SEWRL-030190FMD.

Degoutte, G., and Mathieu, G. (1986). "Experimental Research of Friction Between Soil and Geomembranes or Geotextiles Using a Thirty by Thirty Square Centimeter Shearbox." *Proceedings of the Third International Conference on Geotextiles*, Vienna, Austria, 1251-1256.

Druschel, S.J., and Underwood, E.R. (1993). "Design of Lining and Cover System Sideslopes." *Proceedings of the Geosynthetics '93 Conference*, Vancouver, Canada, Feb, 1341-1355.

Duncan, J.M. (1992). "State-of-the-Art: Static Stability and Deformation Analysis." *Proceedings of Stability and Performance of Slopes and Embankments - II*, ASCE Geotechnical Special Publication, Berkeley, CA, 31(1), 222-266.

Edil, T.B., Ranguette, V.J., and Wuellner, W.W. (1990). "Settlement of Municipal Refuse." *ASTM STP 1070, Geotechnics of Waste Fills - Theory and Practice*, A.O. Landva and G. D. Knowles, Eds., American Society for Testing and Materials, Philadelphia, PA, 225-239.

Eigenbrod, K.D., and Locker, J.G. (1987). "Determination of Friction Values for the Design of Side Slopes Lined or Protected with Geosynthetics." *Canadian Geotechnical Journal*, 24(4), 509-519.

EPRI (1984). "*Comparison of Two Groundwater Flow Models - UNSAT1D and HELP.*" Electric Power Research Institute Topical Report, EPRI CS-3695, Project 1406-1, Oct.

Fassett, J.B., Leonards, G.A., and Repetto, P.C. (1994). "Geotechnical Properties of Municipal Solid Wastes and Their Use in Landfill Design." *Proceedings of the Waste Tech '94 Conference*, National Solid Waste Management Association, Charleston, SC, Jan.

Fenn, D.G., Haney, K.J., and DeGeare, T.V. (1975). "*Use of the Water Balance Method for Predicting Leachate Generation for Solid Waste Disposal Sites.*" U.S. Environmental Protection Agency, Office of Research and Development, Washington, D.C., Report No. EPA/530/SW-169.

Field, C.R., and Nangunoori, R.K. (1994). "Case Study-Efficacy of the HELP Model: A Myth or Reality?" *Proceedings of the Waste Tech '94 Conference*, National Solid Wastes Management Association, Charleston, SC, Jan.

Fourie, A.B., and Fabian, K.J. (1987). "Laboratory Determination of Clay-Geotextile Interaction." *Geotextiles and Geomembranes*, 6(4), 275-294.

Gaind, K.J., and Char, A.N.R. (1983). "Reinforced Soil Beams." *Journal of Geotechnical Engineering*, ASCE, 109(7), 977-982.

Gibson, R.E., and Lo, K.Y. (1961). "A Theory of Soils Exhibiting Secondary Compression." *Acta Polytechnica Scandinavica*, C, 10, 296, 1-15.

Gilbert, R.B., Liu, C.N., Wright, S.G., and Trautwein, S.J. (1995). "A Double Shear Test Method for Measuring Interface Strength." *Proceedings of the Geosynthetics '95 Conference*, Nashville, TN, Feb, 1017-1029.

Giroud, J.P. (1984). "Analysis of Stresses and Elongations in Geomembranes." *Proceedings of the International Conference on Geomembranes*, Denver, CO, 2, 481-486.

Giroud, J.P., and Bonaparte, R. (1989a). "Leakage Through Liners Constructed with Geomembranes, Part I: Geomembrane Liners." *Geotextiles and Geomembranes*, 8(1), 27-67.

Giroud, J.P., and Bonaparte, R. (1989b). "Leakage Through Liners Constructed with Geomembranes, Part II: Composite Liners." *Geotextiles and Geomembranes*, 8(2), 71-111.

Giroud, J.P., Bonaparte, R., Beech, J.F., and Gross, B.A. (1990). "Design of Soil Layer-Geosynthetic System Overlapping Voids." *Geotextiles and Geomembranes*, 9(1), 11-50.

Giroud, J.P., Gross, B.A., and Darrasse, J. (1992). "*Flow in Leachate Collection Layers.*" GeoSyntec Consultants, Internal Report, 62 p.

Giroud, J.P., Badu-Tweneboah, K., and Soderman, K.L. (1994). "Evaluation of Landfill Liners." *Proceedings of the Fifth International Conference on Geotextiles, Geomembranes, and Related Products*, Singapore, 3, 981-986.

Giroud, J.P., and Houlihan, M.F. (1995). "Design of Leachate Collection Layers." *Proceedings of the Fifth International Landfill Symposium*, Sardinia, Oct, in press.

Giroud, J.P., Williams, N.D., and Pelte, T. (1995). "Stability of Geosynthetic-Soil Layered Systems on Slopes." *Geosynthetics International*, 2(5), in press.

Goldman, S.J., Jackson, K., and Burszlynsky, T.A. (1986). "*Erosion and Sediment Control Handbook.*" McGraw-Hill Book Company, New York, Chapter 6, 45 p.

Harder, L.S., Jr. (1991). "Performance of Earth Dams During the Loma Prieta Earthquake." *Proceedings of the Second International Conference on Recent Advances in Geotechnical Earthquake Engineering and Soil Dynamics*, Saint Louis, MO, Mar, 11-15.

Idriss, I.M. (1990). "Response of Soft Soil Sites During Earthquakes." *Proceedings of Symposium to Honor Professor H.B. Seed*, Berkeley, CA.

Idriss, I.M., and Sun, J.I. (1992). "*User's Manual for SHAKE91.*" Center for Geotechnical Modeling, Department of Civil and Environmental Engineering, University of California, Davis, CA.

Ingold, T.S. (1982). "Some Observations on the Laboratory Measurement of Soil-Geotextile Bond." *Geotechnical Testing Journal*, 5(3/4), 57-67.

Jang, D.-J., and Montero, C. (1993). "Design of Liner Systems Under Vertical Expansions: An Alternative to Geogrids." *Proceeding of the Geosynthetics '93 Conference*, Vancouver, Canada, Feb, 1497-1510.

Kavazanjian, E., Jr., Hushmand, B., and Martin, G.R. (1991). "Frictional Base Isolation Using a Layered Soil-Synthetic Liner System." *Proceedings of the Third U.S. Conference on Lifeline Earthquake Engineering*, Technical Council on Lifeline Earthquake Engineering Monograph No. 4, Los Angeles, CA, 1140-1151.

Kavazanjian, E., Jr., Matasović, N., Bonaparte, R., and Schmertmann, G.R. (1995). "Evaluation of MSW Properties for Seismic Analysis." *Proceedings of Geoenvironment 2000*, ASCE Specialty Conference, New Orleans, LA, Feb, 1126-1141.

Kavazanjian, E., Jr., and Matasović, N. (1995). "Seismic Analysis of Solid Waste Landfills." *Proceedings of Geoenvironment 2000*, ASCE Specialty Conference, New Orleans, LA, Feb, 1066-1080.

Khire, M.V., Benson, C.H., and Bosscher, P.J. (1994). "*Final Cover Hydrologic Evaluation- Phase III.*" Environmental Geotechnics Report No. 94-4, University of Wisconsin, Madison, WI, Dec, 142 p.

Kmet, P. (1982). "*EPA's 1975 Water Balance Method-Its Use and Limitations.*" Wisconsin Dept. of Natural Resources, Oct.

Koerner, R.M., and Hwu, B.L. (1991). "Stability and Tension Considerations Regarding Cover Soils on Geomembrane Lined Slopes." *Geotextiles and Geomembranes*, 10, 335-355.

LaGatta, M.D. (1992). "*Hydraulic Conductivity Tests on Geosynthetic Clay Liners Subjected to Differential Settlement.*" MSCE Thesis, University of Texas at Austin.

Landva, A.O., and Clark, J.I. (1990). "Geotechnics of Waste Fill." *ASTM STP 1070, Geotechnics of Waste Fills-Theory and Practice*, A.O. Landva and G.D. Knowles, Eds., American Society for Testing and Materials, Philadelphia, PA, 86-103.

Lane, D.T., Benson, C.H., and Bosscher, P.J. (1992). "*Hydrologic Observations and Modeling Assessments of Landfill Covers.*" Environmental Geotechnics Report No. 92-10, University of Wisconsin, Madison, WI, Dec, 406 p.

Leonards, G.A., and Narain, J. (1963). "Flexibility of Clay and Cracking of Earth Dams." *Journal of the Soil Mechanics and Foundations Division*, ASCE, 89(SM2), 47-98.

Long, J.H., Daly, J.J., and Gilbert, R.B. (1993). "*Structural Integrity of Geosynthetic Lining and Cover Systems for Solid Waste Landfills.*" Department of Civil Engineering, University of Illinois, Project No. OSWR 06-005, Jul.

Long, J.H., Gilbert, R.B., and Daly, J.J. (1994). "Geosynthetic Loads in Landfill Slopes: Displacement Compatibility." *Journal of Geotechnical Engineering*, ASCE, 120(11), 2009-2025.

Makdisi, F.I., and Seed, H.B. (1978). "Simplified Procedure for Estimating Dam and Embankment Earthquake-Induced Deformations." *Journal of Geotechnical Engineering*, ASCE, 104(GT7), 849-867.

Martin, J.P., Koerner, R.M., and Whitty, J.E. (1984). "Experimental Friction Evaluation of Slippage Between Geomembranes, Geotextiles, and Soils." *Proceedings of the International Conference on Geomembranes*, Denver, CO, Industrial Fabrics Association International, 191-196.

Matasović, N. (1991). "Selection of Method for Seismic Slope Stability Analysis." *Proceedings of the Second International Conference on Recent Advances in Geotechnical Earthquake Engineering and Soil Dynamics*, St. Louis, MO, 2, 1057-1062.

Miyamori, T., Iwai, S., and Makiuchi, K. (1986). "Frictional Characteristics of Non-Woven Fabrics." *Proceedings of the Third International Conference on Geotextiles*, Vienna, Austria, 701-705.

Mitchell, R.J., and Mitchell, J.K. (1992). "Stability Evaluation of Waste Landfills." *Stability and Performance of Slopes and Embankments-II*, ASCE Geotechnical Special Publication, 31(2), 1152-1187.

Moore, C.A. (1983). *"Landfill and Surface Impoundment Performance Evaluation."* U.S. Environmental Protection Agency, Washington, D.C., Report No. EPA/530/SW-869-C, 69 p.

Myles, B. (1982). "Assessment of Soil Fabric Friction by Means of Shear Evaluation." *Proceedings of the Second International Conference on Geotextiles,* Las Vegas, NV, 787-791.

Nagasawa, T., and Umeda, Y. (1985). "Effects of the Freeze-Thaw Process on Soil Structure." *Ground Freezing 1985,* Balkema, Rotterdam, 2, 219-225.

Nataraj, M.S., Magauti, R.S., and McManis, K.L. (1995). "Interface Frictional Characteristics of Geosynthetics." *Proceedings of the Geosynthetics '95 Conference,* Nashville, TN, Feb, 1057-1069.

NAVFAC (1983). *"Soil Dynamics, Deep Stabilization, and Special Geotechnical Construction, Design Manual 7.3."* Apr, 77-78.

Newmark, N.M. (1965). "Effects of Earthquakes on Dams and Embankments." *Geotechnique,* 15(2), 139-160.

Pasqualini, E., Roccato, M., and Sani, D. (1993). "Shear Resistance of the Interfaces of Composite Liners." *Proceedings of the Sardinia '93 Fourth International Landfill Symposium,* Cagliari, Italy, Oct, 1457-1471.

Peters, N., Warner, R.S., Coates, A.L., Logsdon, D.S., and Grube, W.E. (1986). "Applicability of the HELP Model in Multilayer Cover Design: A Field Verification and Modeling Assessment." *Land Disposal of Hazardous Waste-Proceedings of the 1986 Research Symposium,* EPA, Cincinnati, OH.

Peyton, R.L., and Schroeder, P.R. (1988). "Field Verification of HELP Model for Landfills." *Journal of Environmental Engineering,* ASCE, 114(2), 247-269.

Peyton, R.L., and Schroeder, P.R. (1993). "Water Balance for Landfills." *Geotechnical Practice for Waste Disposal,* D.E. Daniel, Ed., Chapman & Hall, London, 214-243.

Poorooshasb, H.B. (1991). "Load Settlement Response of a Compacted Fill Layer Supported by a Geosynthetic Overlying a Void." *Geotextiles and Geomembranes,* 10(3), 179-201.

Ritchie, J.T. (1972). "A Model for Predicting Evaporation from a Row Crop with Incomplete Cover." *Water Resources Research,* 8(5), 1204-1213.

Sagaseta, C. (1987). "Analysis of Undrained Soil Deformation Due to Ground Loss." *Geotechnique*, 37(3), 301-320.

Saxena, S.K., and Wong, Y.T. (1984). "Frictional Characteristics of a Geomembrane." *Proceedings of the International Conference on Geomembranes*, Denver, CO, Industrial Fabrics Association International, 187-190.

Scharch, P.E. (1985). "*Water Balance Analysis Program for: the IBM-PC Micro-Computer*." State of Wisconsin Department of Natural Resources, May.

Schnabel, P.B., Lysmer, J., and Seed, H.B. (1972). "*SHAKE: A Computer Program for Earthquake Response Analysis of Horizontally Layered Sites*." Report No. EERC 72-12, Earthquake Engineering Research Center, University of California, Berkeley, CA.

Schroeder, P.R., Lloyd, C.M., and Zappi, P.A. (1994a). "*The Hydrologic Evaluation of Landfill Performance (HELP) Model, User's Guide for Version 3*." U.S. Environmental Protection Agency, Office of Research and Development, Washington, D.C., Report No. EPA/600/R-94/168a, Sep.

Schroeder, P.R., Dozier, T.S., Zappi, P.A., McEnroe, B.M., Sjostrom, J.W., and Peyton, R.L. (1994b). "*The Hydrologic Evaluation of Landfill Performance (HELP) Model Engineering Documentation for Version 3*." U.S. Environmental Protection Agency, Office of Research and Development, Washington, D.C., Report No. EPA/600/R-94/168b, Sep, 116 p.

Seed, R.B. (1989). "*Final Results of Direct Shear Tests; Liner Interface Strength, Cedar Hills Regional Landfill*." prepared for CH2M Hill.

Seed, H.B., and Idriss, I.M. (1982). "Ground Motions and Soil Liquefaction During Earthquakes." *Monograph No. 5*, Earthquake Engineering Research Institute, Berkeley, CA, 134 p.

Seed, R.B., Mitchell, J.K., and Seed, H.B. (1988). "*Slope Stability Failure Investigation: Landfill Unit B-19, Phase I-A, Kettleman Hills, California*." University of California at Berkeley, Berkeley, CA, Report No. UCB/GT/88-01.

Seed, R.B., and Boulanger, R.W. (1991). "Smooth HDPE-Clay Liner Interface Shear Strengths: Compaction Effects." *Journal of Geotechnical Engineering*, ASCE, 117(4), 686-693.

Seed, R.B., and Bonaparte, R. (1992). "Seismic Analysis and Design of Lined Waste Fills: Current Practice." *Proceedings of Stability and Performance of Slopes and Embankments - II*." ASCE Geotechnical Special Publication, Berkeley, CA, 31(2) 1521-1545.

Sharma, S. (1994). *"XSTABL, Reference Manual, Version 5."* Interactive Software Designs, Inc., Moscow, ID.

Sharma, H.D., Dukes, M.T., and Olsen, D.M. (1990). "Field Measurements of Dynamic Moduli and Poisson's Ratios of Refuse and Underlying Soils at a Landfill Site." *ASTM STP 1070, Geotechnics of Waste Landfills -Theory and Practice*, A.O. Landva and G.D. Knowles, Eds., American Society for Testing and Materials, Philadelphia, PA, 57-70.

Siegel, R.A. (1975). *"STABL User Manual."* Report No. JHRP-75-9, School of Civil Engineering, Purdue University, West Lafayette, IN.

Singh, S., and Sun, J.I. (1995). "Seismic Evaluation of Municipal Solid Waste Landfills." *Proceedings of Geoenvironment 2000*, ASCE Specialty Conference, New Orleans, LA, Feb, 1081-1096.

Sowers, G.F. (1973). "Settlement of Waste Disposal Fills." *Proceedings of the Eighth International Conference on Soil Mechanics and Foundation Engineering*, Moscow, 207-210.

Stark, T.D., and Poeppel, A.R. (1994). "Landfill Liner Interface Strengths from Torsional-Ring-Shear Tests." *Journal of Geotechnical Engineering*, ASCE, 120(3), 597-615.

Stulgis, R.P., Soydemic, C., and Telegener, R.J. (1995). "Predicting Landfill Settlement." *Proceedings of Geoenvironment 2000*, ASCE Specialty Conference, New Orleans, LA, Feb, 980-993.

Swan, R.H., Bonaparte, R., Bachus, R.C., Rivette, C.A., and Spikula, D.R. (1991). "Effect of Soil Compaction Conditions on Geomembrane-Soil Interface Strength." *Geotextiles and Geomembranes*, 10, 523-529.

Thornthwaite, C.W., and Mather, J.R. (1957). "Instructions and Tables for Computing Evapotranspiration and the Water Balance." Drexel University of Technology, *Publications in Climatology*, X(3).

Udoh, F.D. (1991). *"Minimization of Infiltration Into Mining Stockpiles Using Low Permeability Covers."* Dissertation Proposal, Dept. of Materials Science and Engineering, Mining Engineering Program, University of Wisconsin, Madison, WI.

USACOE (1970). *"Engineering and Design Stability of Earth and Rock-Fill Dams."* U.S. Army Corps of Engineers, Manual EM 1110-2-1902, Apr.

USACOE (1990). *"HEC-1 Flood Hydrograph Package, User's Manual."* Computer Program Document No. 1A, U.S. Army Corps of Engineers, Davis, CA.

USDA-SCS (1985). "*User's Guide for the CREAMS Computer Model.*" Washington Computer Center Version, Technical Release 72.

USDA-SCS (1986). "*Urban Hydrology for Small Watersheds: Technical Release 55.*" U.S. Department of Agriculture, Jun.

USEPA (1991). "*Design and Construction of RCRA/CERCLA Final Covers.*" U.S. Environmental Protection Agency Office of Research and Development, Washington, D.C., Report No. EPA/625/4-91/025, May.

USEPA (1992). "*Federal Register.*" U.S. Environmental Protection Agency, Washington, D.C., 57(124), 26 Jun, 28626-28628.

USEPA (1995). "*RCRA Subtitle D (258) Seismic Design Guidance for Municipal Solid Waste Landfill Facilities.*" U.S. Environmental Protection Agency, Office of Research and Development, Washington, D.C., Report No. EPA/600/R-95/051, Apr, 143 p.

USGS (1982). "*Probabilistic Estimates of Maximum Acceleration and Velocity in Rock in the Continuous United States.*" United States Geological Survey, Open-File Report 82-1033, 99 p.

USGS (1990). "*Probabilistic Earthquake Acceleration and Velocity Maps for the United States and Puerto Rico.*" U.S. Geological Survey Miscellaneous Field Studies Map MF-2120, Scale 1:7,500,000.

Wall, D.K., and Zeiss, C. (1995). "Municipal Landfill Biodegradation and Settlement." *Journal of Environmental Engineering*, ASCE, 121(3), 214-224.

Williams, J.R., and Hann, R.W. (1978). "*Optimal Operation of Large Agricultural Watersheds with Water Quality Constraints.*" Texas A&M University, Texas Water Resources Institute, TR-96.

Williams, N.D., and Houlihan, M.F. (1987). "Evaluation of Interface Friction Properties Between Geosynthetics and Soils." *Proceedings of the Geosynthetics '87 Conference*, New Orleans, LA, 2, 616-627.

Wilson-Fahmy, R.F., and Koerner, R.M. (1992). "*Stability Analysis of Multi-Lined Slopes in Landfill Applications.*" Geosynthetic Research Institute, Drexel University, GRI Report No. 8, Dec.

Wilson-Fahmy, R.F., and Koerner, R.M. (1993). "Finite Element Analysis of Stability of Cover Soil on Geomembrane-Lined Slopes." *Proceedings of the Geosynthetics '93 Conference*, Vancouver, Canada, Feb, 1425-1438.

Wright, S.G. (1991). "*UTEXAS3 - A Computer Program for Slope Stability Calculations.*" Austin, Texas, Sep 1991.

Yegian, M.K., and Lahlaf, A.M. (1992). "Dynamic Interface Shear Strength Properties of Geomembranes and Geotextiles." *Journal of Geotechnical Engineering,* ASCE, 118(5), 760-779.

Yegian, M.K., Yee, Z.Y., and Harb, J.N. (1995). "Seismic Response of Geosynthetic/Soil Systems." *Proceedings of Geoenvironment 2000,* ASCE Specialty Conference, New Orleans, LA, Feb, 1113-1125.

LANDFILL GAS AND GROUNDWATER CONTAMINATION

Richard Prosser[1], P.E.
Alan Janechek[2], P.E.

ABSTRACT

As landfill gas (LFG) migrates from a landfill, organic contaminants travel with it. These contaminants, commonly referred to as volatile organic compounds (VOCs), have been known to migrate to underlying groundwater. This paper looks at the mechanisms for contamination by LFG constituents and means of controlling it. The four basic mechanisms that cause contamination are:

1. Direct contact of groundwater with LFG,
2. Formation of landfill gas condensate water in the soil outside of a landfill
3. LFG contamination of the vadose zone and infiltration water carrying the VOCs to the groundwater
4. Leachate leaving the landfill and migrating to the groundwater.

Of these four mechanisms, this paper will look at the contamination caused by the first three. These represent methods of groundwater contamination by LFG. Leachate contamination is not within the scope of this paper.

INTRODUCTION

The Resource Conservation and Recovery Act (RCRA) requires that landfill gas (LFG) be controlled so that the concentration of methane at a landfill's property line is less than 5% by volume *(40 CFR Part 258.23)*. The purpose of this requirement is to protect adjacent properties from explosive conditions. This requirement does not address the need to control migration due to the presence of Volatile Organic

---

[1]President, Gas Control Engineering, 1205 N. Red Gum St., Suite B, Anaheim, CA 92806
[2]Project Manager, Gas Control Engineering, 1205 N. Red Gum St., Suite B, Anaheim, CA 92806

Compounds (VOCs) in the LFG. The migration of VOCs can cause several problems. First, they can travel to groundwater causing contamination; second, they can infiltrate adjacent structures and potentially cause health risk problems; and third, they can cause odor problems. This paper focuses exclusively on contamination of groundwater by VOCs.

BACKGROUND

LFG testing performed on a number of California landfills demonstrates that most LFG contains some VOCs (3, 4). These contaminants include organic acids, chlorinated hydrocarbons, and numerous other hydrocarbons. The contaminants of greatest concern are typically the chlorinated hydrocarbons. This is because many of the chlorinated hydrocarbons are considered a health risk at low concentrations and because they are not as readily decomposed in the soil by naturally occurring aerobic bacteria. The subsequent movement of these constituents to groundwater can result in concentrations exceeding State or Federal drinking water standards.

Most of the landfills in the U.S. are affected by the presence of VOCs in the underlying groundwater. In a recent study (3) of the results of monitoring RCRA Appendix IX (40 CFR parts 264 and 265) compounds at 479 disposal sites, 84 percent of the detectable compounds were VOCs. The most commonly occurring compounds were the chlorinated hydrocarbons, including the 8 most frequently detected compounds. Methylene chloride was the most commonly detected compound followed closely by trichloroethene and tetrachloroethene. Each of these compounds was detected at over 20% of the landfill sites tested. Because LFG frequently emanates from a landfill into the surrounding soils, there is a possibility that LFG was either the primary source of contamination or a strong contributor. The reason for this assertion is discussed later in this paper.

POTENTIAL FOR CONTAMINATION

Because VOC concentrations vary widely in each landfill, establishing a minimum control level for LFG at the perimeter based on protection of groundwater does not seem practical. (Note: Protection of human health in adjacent structures from VOCs in LFG vapor may be a more significant concern than groundwater at some sites.) Additional factors that can affect the potential for groundwater contamination include LFG generation rates, liner and formation permeabilities, distance to groundwater, and VOC attenuation by soil bacteria. To quantify the mass of contamination that can be discharged from a landfill before experiencing significant groundwater contamination (defined as levels of VOCs in groundwater at or above the MCL) is difficult. Each site is unique and needs to be treated accordingly. In addition, by the time contamination is detected in groundwater monitoring wells a significant amount of VOC mass will likely have accumulated in the unsaturated zone.

The best way to demonstrate the potential for LFG contamination is to describe a typical scenario using several assumptions. Assume a 162,000 sq. meter (40-acre) landfill with 908×10^6 Kg (1,000,000 tons) of refuse in place with an average LFG generation rate of 279 Nm³/Hr (0.25 MM cubic feet per day)[1]. (The LFG generation rate is equivalent to an annual LFG yield of 0.00265 Nm³/Kg (0.045 cubic feet per pound of refuse).) The gas is assumed to contain 10 ppmv trichloroethene (TCE) as its only contaminant. (TCE was selected because it is frequently found in groundwater (3).) Assuming 10% of the LFG with its VOC fraction goes into the vadose zone and 10% of this eventually reaches groundwater. The magnitude of contamination of trichloroethene VOC is

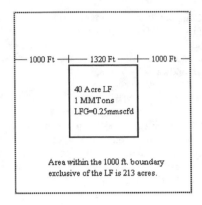

Figure 1: Example Landfill With 1000 Ft Border

calculated by determining the resulting volume of groundwater that would be effected by the contamination. The potential contaminated water volume is estimated by limiting the average contamination in the water to the MCL for TCE. The procedure used in Equation 1 is to first determine the lbs. of TCE in groundwater and divide this by the MCL of TCE to yield the water contamination volume.

$$V = (0.1)(0.1)\left(\frac{10\ ppmv}{1 \times 10^6}\right)\left(\frac{279\ \frac{Nm^3}{Hr.} \times 24\ \frac{Hr.}{Day}}{\frac{22.4\ L}{g\text{-}mole} \times \frac{1\ m^3}{1000\ L}}\right)\left(\frac{131\ g\text{-}TCE}{g\text{-}mole}\right)\left(\frac{L}{5\mu g}\right)\left(\frac{1 \times 10^6\ \mu g}{g}\right)\left(\frac{365\ Days}{Yr.}\right) \quad (1)$$

$$= 286 \times 10^6\ \frac{L}{Yr.} \quad (75 \times 10^6\ gallons/Yr.)$$

where;

V	= Contaminated Water Volume at the MCL
10 ppmv	= assumed volume concentration of TCE in LFG
5 µg/L	= drinking water Maximum Contaminant Level for TCE (5 ppbw)

---

[1] Standard conditions used in calculations are 0°C and 760 mm Hg for metric units and 60°F and 14.7 psia for English units.

This annual volume of contaminated water would cover an area of approximately 142,000 m² (115 acres), assuming a 3.05 m (10 foot) thickness of uniformly contaminated aquifer and a 0.20 soil porosity.

Landfill liners provide only a partial barrier to LFG movement. Clay-lined landfills are subject to the same convective and diffusive mechanisms as an unlined landfill. Geomembrane liners are subject to construction deficiencies and deterioration. As an example of the magnitude of potential migration through a clay liner, consider the example of only diffusive transport of TCE through a typical landfill liner (1).

$$Q = \frac{AD\alpha^{4/3}C_i - C_w}{L} \tag{2}$$

where;

Q	= Diffusive flow of TCE
A	= Area of landfill bottom (40 acre landfill)
D	= Diffusion coefficient of TCE at 20°C
$\alpha^{4/3}$	= Gas filled porosity of liner (assumed no saturation)
C_i	= Concentration of TCE above liner (10 ppmv)
C_w	= Concentration of TCE below liner (9.99 ppmv)
L	= Depth of liner

$$Q_{TCE} = \frac{(162,000 \text{ m}^2)\left(\frac{10,000 \text{ cm}^2}{\text{m}^2}\right)\left(\frac{0.067 \text{ cm}^2}{\text{sec.}}\right)(0.20)^{4/3}\left(\frac{10 - 9.99}{1,000,000}\right)\left(\frac{131 \text{ g}}{\text{g-mole}}\right)\left(\frac{1 \text{ g-mole}}{22.4 \text{ L}}\right)\left(\frac{1 \text{ L}}{1000 \text{ cm}^3}\right)\left(\frac{86,400 \text{ sec.}}{\text{day}}\right)}{61 \text{ cm}} \tag{3}$$

$$= 1.05 \text{ g/day TCE}$$

and

$$V = \left(1.05\frac{\text{gTCE}}{\text{Day}}\right)\left(\frac{365 \text{ Days}}{\text{Yr.}}\right)\left(\frac{1}{\frac{5 \mu g}{L}} \times \frac{1 \times 10^6 \mu g}{1 g}\right)$$

$$= 77 \times 10^6 \text{ L/Yr. } (20 \times 10^6 \text{ gallons/Yr.})$$

LFG CONTAMINATION MECHANISMS

The three basic mechanisms which cause groundwater contamination by VOCs in LFG within the vadose zone are described briefly in this section and are shown on **Figure 2**, LFG Contamination Mechanisms.

1. Direct contact with groundwater by LFG.

LFG migrating from a site tends to travel the path of least resistance. Generally that path is towards the atmosphere. Two dimensional modeling also shows that LFG migrates downward prior to escaping to the atmosphere. As LFG reaches the capillary zone, the VOCs in the LFG have a good opportunity to be absorbed into the groundwater.

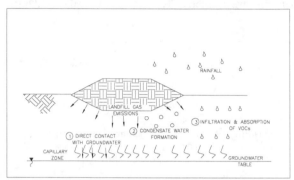

Figure 2: LFG Contamination Mechanisms

2. Formation of landfill gas condensate in the soil adjacent to the landfill.

Landfill gas temperatures typically range from 27 to 38° C (80 to 100° F) within a landfill. As LFG moves through refuse, it is typically saturated with water at these temperatures. The soil surrounding a landfill is usually cooler, and for discussion purposes is assumed to be 16° C (60° F). The difference in these two conditions will result in the formation of condensate water outside the refuse mass. Partitioning of VOCs from the vapor to liquid phase will typically result in trace concentrations of VOCs within the condensate.

3. LFG contamination of the vadose zone and infiltration water carrying the VOCs to the groundwater

As LFG migrates through the soil adjacent to a landfill, VOC mass adheres to the soil matrix in either vapor or liquid phase. As rainwater, irrigation or other water infiltrates from above, it reaches equilibrium with the VOCs present in the soil. Provided these VOCs are not consumed by bacteria or stripped from the water, they may eventually reach groundwater.

Henry's Law

The three mechanisms described above are all based on absorption of the VOCs from the vapor phase to the liquid phase. The magnitude of the equilibrium absorption and the resulting relative equilibrium concentrations of the VOCs in the air and water phases can be calculated using Henry's Law. Henry's law is expressed as follows:

$$X_{AQ} = \frac{P_x}{K} \tag{4}$$

where;

X_{AQ} = Aqueous concentration of the gas
K = Henry's Law constant at specified temperature
P_x = Partial pressure of the gas

To demonstrate the partitioning of VOC between the water phase and the vapor phase, LFG is assumed to contain trichloroethene, benzene and vinyl chloride as the only contaminants. The assumed gas phase concentration of each is 1 ppmv and the absorption of each does not hinder the absorption of the others.

A. Benzene Concentration in Water:

$$X_B = \frac{1 \times 10^{-6}(1 \text{ atm} - 0.0313 \text{ atm})(78g/g*mol)(1 \times 10^6 \mu g/g)}{(5.43 \times 10^{-3} m^3 atm/g*mol)(1000L/m^3)} \tag{5}$$

= 13.9 µg/L Benzene   (Note: MCL for Benzene = 1 µg/L)

where;

K   = 5.43x10^{-3} m^{3}atm/g*mol is the Henry's Law constant for Benzene at 20°C
78 g/g*mol = the molecular weight of Benzene
0.0313 atm = the correction for the vapor pressure of water

B. TCE Concentration in Water:

$$X_{TCE} = \frac{1x10^{-6}(1 \text{ atm} - 0.0313 \text{ atm})(131.4g/g*mol)(1x10^6 \mu g/g)}{(1.03x10^{-2} m^3 atm/g*mol)(1000L/m^3)} \tag{6}$$

= 12.3 µg/L TCE   (Note: MCL for TCE = 5 µg/L)

C. Vinyl chloride Concentration in Water:

$$X_{VC} = \frac{1\times10^{-6}(1 \text{ atm} - 0.0313 \text{ atm})(62.5 \text{ g}/\text{g}*\text{mol})(1\times10^{6} \mu g/\text{g})}{(1.07\times10^{-2} \text{ m}^3 \text{atm}/\text{g}*\text{mol})(1000\text{L}/\text{m}^3)} \tag{7}$$

$$= 5.8 \text{ μg/L Vinyl Chloride} \qquad \text{(Note: MCL for VC} = 0.5 \text{ μg/L)}$$

LANDFILL GAS GENERATION

Landfill gas is generated by the decomposition of refuse by anaerobic bacteria. A simple equation for the production of LFG from cellulose is as follows:

$$\overset{\text{bacteria}}{C_6H_{10}O_5 + H_2O \quad \rightarrow \quad 3CH_4 + 3CO_2} \tag{8}$$

Because LFG is primarily generated by the decomposition of solids in the refuse, the resulting gas occupies a substantially larger volume than the refuse. This causes pressure to build within the landfill as the gas tries to expand to its natural volume. The level of pressure developed is a function of the tortuousity and restrictions the gas has to travel through to expand within the refuse. This process creates the pressure for convective movement of LFG in soil.

The magnitude of pressures typically found at the bottom of a landfill have been measured at pressures ranging from low levels of less than one inch of water column to as high as 4 atmospheres (Operating Industries landfill, Monterey Park, California). The conditions that cause high pressures usually include deep and wet refuse and low permeable soil (1×10^{-6} cm/sec or less) surrounding the landfill.

If LFG is under pressure, it can migrate great distances due to convective transport. Circumstances that allow migration often include highly permeable strata sandwiched between less permeable material. This can also include the sand bedding used to backfill utility pipes and other man-made fills. Many sites have shown measurable concentrations of the constituents of LFG as much as a thousand feet from the site.

MAGNITUDE OF CONTAMINATION MECHANISMS

In order to design a control strategy for minimizing the amount of VOC mass migrating from a landfill, the unique characteristics of each site must be considered. However, it is useful to look at the relative contributions from each of the previously described mechanisms. Evaluation of the potential for VOC migration from condensate formation and vapor phase transport are discussed below.

Condensate Water Formation

The amount of water condensing from LFG is primarily a function of temperature differences. A simple equation used to calculate the amount of water that is present in LFG is based on the partial pressure of the water.

$$H_2O\% = \left(\frac{ppH_2O}{P_T}\right) \times 100 \tag{9}$$

where;

ppH_2O = Partial Pressure of the water (Water vapor pressure)
P_T = System total pressure

The amount of water that condenses is calculated by determining the water content of the LFG in the landfill and again within the soil. The difference between these values is equal to the water that condenses. For the example given previously, if 10% of 0.25 mm of dry LFG exits the landfill at 32° C (90° F) and cools to 16° C (60° F) in the soil with a system pressure of 760 mm Hg (14.7 PSIA), the amount of water that condenses is calculated:

$$\text{Volume of Water in LFG (Nm}^3) = \frac{H_2O\% \times LFG~(Nm^3)}{(1 + H_2O\%)} \tag{10}$$

The fraction of water in the landfill gas while it is in the landfill and in the soil is calculated.

$$H_2O\% \text{ at } 32°C = \left(\frac{15.1~\text{mm Hg}}{760~\text{mm Hg}}\right) \times 100 = 2\% \tag{11}$$

$$H_2O\% \text{ at } 16°C = \left(\frac{2.64~\text{mm Hg}}{760~\text{mm Hg}}\right) \times 100 = 0.34\% \tag{12}$$

Therefore water condensate formation ΔW (Nm^3)

$$\Delta W = \left[\frac{0.02 \times (0.1 \times 279~\text{Nm}^3/\text{Hr.}}{(1+0.02)} - \frac{0.0034 \times (0.1 \times 279~\text{Nm}^3/\text{Hr.}}{1+0.0034}\right] \times \left(\frac{24~\text{Hr.}}{\text{Day}}\right) \tag{13}$$

$$\Delta W = 10.86 \frac{\text{Nm}^3~\text{Water Vapor}}{\text{Day}} \rightarrow 8.73 \frac{L}{\text{Day}}$$

Based on previous Henry's Law calculations, the TCE mass present in the condensate water is approximately:

$$TCE = 8.73 \, \frac{L}{Day} \times 12.2 \frac{\mu g}{L} = 107 \frac{\mu g}{Day} \tag{14}$$

The annual volume of groundwater that would be contaminated by this is calculated:

$$V = \left(107 \frac{\mu g}{Day}\right)\left(365 \frac{Days}{Yr.}\right)\left(\frac{L}{5\mu g}\right) \tag{15}$$

$$= 7811 \, \frac{L}{Yr.}$$

This calculation shows that condensate water does not appear to be a major contamination mechanism. Compared to the potential to contaminate calculation (Equation 1), this is a small quantity.

Vapor Phase Contamination

Vapor phase contamination occurs by two processes. These are direct contact of LFG with groundwater and leaching of VOCs from the vadose zone by infiltrating water. Because of the potential magnitude of VOCs in the soil, these combined mechanisms have the opportunity to cause the most water contamination. Both of these mechanisms are enhanced by the tendency of LFG to travel both vertically and laterally away from the landfill. By distributing the VOCs into a larger area there is increased opportunity for water infiltration or VOC diffusion to groundwater to cause contamination.

Consider the sample landfill shown on **Figure 1**. If we assume that VOCs migrate laterally due to convective transport, the areal extent of contamination outside the landfill is estimated to be 213 acres. The 1000 ft. border is selected because this is a commonly established regulatory guideline for LFG emissions.

The contribution of each mechanism is not quantified because these would be very specific to the site under consideration. The contribution of each could range from small to large depending on the water infiltration rate, proximity to groundwater and other geologic factors and conditions. The combination of these processes may contribute significantly to future water contamination even if the contamination source is controlled. This is because of residual VOC's in the vadose zone that can migrate to groundwater.

PROTECTION OF GROUNDWATER

Protecting groundwater from LFG contamination is best achieved by containing LFG within the landfill. Landfills often times have gas control systems present within them. Taking advantage of these systems to help protect groundwater is not a common function. A more common practice is to control methane gas at the landfill's perimeter. This type of approach misses a great opportunity to help protect the environment and reduce the potential of future groundwater clean-up.

The scope of this paper does not include presenting operational and design procedures for minimizing the amount of VOCs escaping from the landfill. Mitigation of VOCs once they reach the unsaturated zone beneath a landfill are discussed briefly. Two methods for reducing the capacity of VOCs to migrate to groundwater from the unsaturated zone are as follows:

1. Increase Volatilization of VOCs in the Unsaturated Zone

The process of removing VOCs from soil has been used extensively to remediate contaminated soils. This process flushes the soil with stripping air to evaporate the VOCs. This process is also supplemented by bacterial degradation of the VOCs in an aerobic environment. Not all VOCs are easily decomposed by aerobic bacterial. For these stripping will be the primary mechanism for removal.

The process commonly used to strip VOC's from the soil is the installation of gas extraction wells. Wells may be placed inside or outside the refuse mass. Because of the potential for wells located outside the refuse to pull VOCs from the landfill however, these need to be used cautiously so as not to increase the contamination area.

Using wells within the refuse is a preferred technique. In addition to controlling LFG emissions, this process can also be used to remove VOCs from the vadose zone. The difficulty with this is preventing landfill fires which result from excessive gas extraction rates.

2. Insitu Degradation of VOCs in the Unsaturated Zone

The second method to reduce VOCs is to degrade them with bacteria. Methaotrophic degradation of chlorinated hydrocarbons can be reasonably effective. In this process, the methane present in the landfill gas is used as a fuel source. Enzymes excreted by the bacteria can co-metabolize the chlorinated hydrocarbons (2). The difficulty with this process is making sure sufficient methane and oxygen are present to oxidize residual VOCs. Operation of perimeter wells will create an aerobic environment. This is suitable for bacterial decomposition of most non-chlorinated hydrocarbons. However, chlorinated hydrocarbons decompose slowly in an aerobic environment. Ultimately the solution is to prevent VOCs migration into the vadose zone in the first place.

CASE HISTORY: ELSINORE LANDFILL

The Elsinore Landfill, located in Lake Elsinore, CA is an unlined, former Class III (municipal waste) landfill which was closed in 1986. The 44 acre site received approximately 543,000 m³ (710,000 cubic yards) of refuse between 1953 and 1986. The landfill is located in an arid environment with an average annual ambient temperature of 17° C (63° F) and an average annual rainfall of less than 38 cm (15 inches) per year. Average depth to groundwater below the landfill is approximately 15.24 m (50 feet). The landfill was capped with two feet of clay in 1992.

The anticipated volume of leachate generation in this environment is relatively small, and in fact drilling in the refuse mass which took place in 1992 indicated a relatively dry refuse. Ongoing quarterly monitoring at the site does not indicate leachate contamination of the underlying groundwater. A good indicator of leachate contamination, chloride, does not exhibit any statistically significant increase in downgradient concentrations of chloride versus upgradient concentrations.

Groundwater monitoring has indicated a number of volatile organic compounds (VOCs) both upgradient and downgradient of the landfill. Table 1 is a summary of all detected compounds in the groundwater monitoring wells located at the site. Because the surrounding land use is a wildlife habitat, it is probable that the landfill is the source of VOC contamination. The apparent source of contamination is via two mechanisms described previously, condensate formation and migration in the liquid phase or direct contact of LFG constituents with groundwater and subsequent air/liquid phase partitioning. Leachate is not considered a mechanism because of the dry refuse state.

The expected volume of selected VOCs from condensate formation were conservatively calculated based on the known flow rate of LFG, VOC concentrations in LFG, LFG cooling of -1° C (30° F), and complete migration of the condensate to groundwater (no attenuation, degradation, or retardation). The calculated volume of these VOC constituents in the condensate water do not support the measured concentrations in the groundwater.

A landfill gas collection system was installed to reduce methane concentrations in the perimeter probes and to help the groundwater contamination problem. A layout of the landfill and collection system is shown on **Figure 3**.

Following installation of the LFG system the methane concentration in the perimeter probes decreased to acceptable levels. Additionally, the VOC concentration in the groundwater wells showed a slight decrease.

The evidence suggests that the LFG collection system had an impact on the VOC present in the groundwater. This system was able to control the methane concentration at the landfill's perimeter and cause a small reduction in the VOCs concentrations in the groundwater. Because of expected VOCs reservoirs in the vadose zone caused by years of uncontrolled LFG emissions, it is expected to take years to demonstrate VOCs control and protection of groundwater.

REFERENCE SOURCES

1. Manahan, Stanley E.   "Fundamentals of Environmental Chemistry".   Lewis Publishers, 1993.

2. McCarty, Perry L.   Bioengineering Issues Related to In Situ Remediation of Contaminated Soils and Groundwater.   Environmental Biotechnology, Plenum Publishing Corporation, 1988.

3. Plumb Jr., R.H.   "The Occurrence of Appendix IX Organic Constituents in Disposal Site Ground Water".   Ground Water Monitoring and Remediation, Spring 1991, pp. 157 - 164.

4. "The Landfill Testing Program: Data Analysis and Evaluation Guidelines". California Air Pollution Control Officers Association Technical Review Group, Landfill Gas Subcommittee, and California Air Resources Board Summary Source Division, September 13, 1990.

TABLE 1

SUMMARY OF DETECTED COMPOUNDS IN THE GROUNDWATER MONITORING WELLS

Elsinore Sanitary Landfill
Summary of Detected Parameters (1st Quarter 1995)

Well ID: E-1 — Date Sampled: 2/15/95

Parameter	Value	Units
Antimony (Sb)	0.004	mg/l
Arsenic (As)	0.016	mg/l
Barium (Ba)	0.029	mg/l
Bicarbonate(HCO3)	143	mg/l
Boron (B)	0.11	mg/l
Calcium (Ca)	74.1	mg/l
Chloride (Cl)	45.9	mg/l
Copper (Cu)	0.005	mg/l
Fluoride (F)	0.27	mg/l
LAB pH	7.5	units
Magnesium (Mg)	14.3	mg/l
Manganese (Mn)	0.007	mg/l
Mercury (Hg)	0.0003	mg/l
Molybdenum (Mo)	0.014	mg/l
Nickel (Ni)	0.008	mg/l
Nitrate (NO3-N)	10.9	mg/l
Potassium (K)	3.5	mg/l
Selenium (Se)	0.007	mg/l
Sodium (Na)	37.8	mg/l
Specific Conductance	737	umhoic
Sulfate (SO4)	60.7	mg/l
Total Dissolved Solids	398	mg/l
Total Hardness	244	mg/l
Total Organic Carbon (TOC)	2	mg/l
Total Phosphorus (P)	0.058	mg/l
Vanadium (V)	0.051	mg/l
Zinc (Zn)	0.008	mg/l

Well ID: E-2 — Date Sampled: 2/15/95

Parameter	Value	Units
1,1-Dichloroethane	17	ug/l
1,2-Dichlorobenzene	0.13	ug/l
1,4-Dichlorobenzene	0.24	ug/l
Arsenic (As)	0.008	mg/l
Barium (Ba)	0.033	mg/l
Bicarbonate(HCO3)	242	mg/l
Boron (B)	0.095	mg/l
Calcium (Ca)	142	mg/l
Chloride (Cl)	101	mg/l
cis-1,2-Dichloroethane	0.74	ug/l
Copper (Cu)	0.003	mg/l
Dichlorodifluoromethane	17	ug/l
Fluoride (F)	0.29	mg/l
LAB pH	7.4	units
Magnesium (Mg)	19.9	mg/l
Manganese (Mn)	0.005	mg/l
Methylene chloride	0.82	ug/l
Molybdenum (Mo)	0.014	mg/l
Nickel (Ni)	0.008	mg/l
Nitrate (NO3-N)	2.8	mg/l
Potassium (K)	4.1	mg/l
Sodium (Na)	46.6	mg/l
Specific Conductance	1130	umhoic
Sulfate (SO4)	124	mg/l
Tetrachloroethene	6	ug/l
Trichloroethene	0.89	ug/l
Trichlorofluoromethane	0.037	ug/l
Zinc (Zn)	0.006	mg/l

Well ID: E-3 — Date Sampled: 2/15/95

Parameter	Value	Units
1,1-Dichloroethane	6	ug/l
1,1-Dichloroethene	1.1	ug/l
1,2-Dichlorobenzene	0.14	ug/l
1,2-Dichloroethene	0.56	ug/l
1,4-Dichlorobenzene	1.1	ug/l
Arsenic (As)	0.013	mg/l
Barium (Ba)	0.012	mg/l
Benzene	1.3	ug/l
Bicarbonate(HCO3)	265	mg/l
Boron (B)	0.064	mg/l
Calcium (Ca)	171	mg/l
Chemical Oxygen Demand COD	5.9	mg/l
Chloride (Cl)	105	mg/l
Chlorobenzene	0.15	ug/l
cis-1,2-Dichloroethane	2.9	ug/l
Copper (Cu)	0.003	mg/l
Dichlorodifluoromethane	30	ug/l
Fluoride (F)	0.16	mg/l
LAB pH	7.2	units
Magnesium (Mg)	37.2	mg/l
Manganese (Mn)	0.005	mg/l
Methylene chloride	4.6	ug/l
Molybdenum (Mo)	0.005	mg/l
Nickel (Ni)	0.004	mg/l
Nitrate (NO3-N)	1.1	mg/l
Potassium (K)	6.5	mg/l
Sodium (Na)	47.3	mg/l
Specific Conductance	1450	umhoic
Sulfate (SO4)	274	mg/l
Tetrachloroethene	5	mg/l
Total Dissolved Solids	844	mg/l
Total Hardness	582	mg/l
Total Organic Carbon (TOC)	4.3	mg/l
Total Organic Halogens (TOX)	0.047	mg/l
Total Phosphorus (P)	0.027	mg/l
Trichloroethene	3	ug/l
Trichlorofluoromethane	1	ug/l
Vanadium (V)	0.025	mg/l
Vinyl chloride	0.7	ug/l
Zinc (Zn)	0.004	mg/l

* - Exceeds "Maximum Contaminant Level" (MCL) per Title 22 Sections 64435 and 64444.5 or "Drinking Water Action Level" (DWAL) per CA Department of Health Services as Indicated.

Run Date: 5/9/95 2:09:46 PM

Riverside County Waste Resources Management District

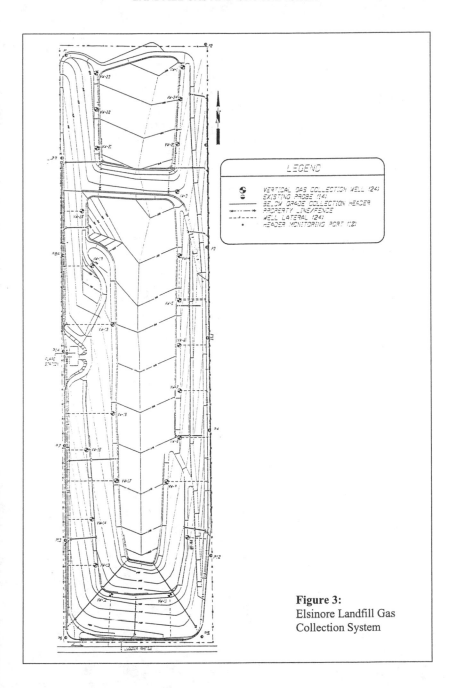

Figure 3:
Elsinore Landfill Gas
Collection System

SUBJECT INDEX
Page number refers to first page of paper

AUTHOR INDEX
Page number refers to first page of paper